RESEARCH REPORT SERIES, INSTITUTE FOR SOCIAL RESEARCH

The Superconducting Super Collider at the Stockbridge, Michigan, Site

Community Support and Land Acquisition

Richard W. Stoffle
Michael W. Traugott
Camilla L. Harshbarger
Florence V. Jensen
Michael J. Evans
Paula Drury

Survey Research Center / Center for Political Studies
Institute for Social Research
The University of Michigan

1988

The Research Report Series of the Institute for Social Research is composed of significant reports published at the completion of a research project. These reports are generally prepared by the principal research investigators and are directed to selected users of this information. Research Reports are intended as technical documents which provide rapid dissemination of new knowledge resulting from ISR research.

Library of Congress Cataloging-in-Publication Data

The Superconducting super collider at the Stockbridge,
Michigan, site.

(Research report series)
Bibliography: p.
1. Superconducting Super Collider--Location.
2. Stockbridge (Mich.) I. Stoffle, Richard W.
II. University of Michigan. Institute for Social
Research. III. Series: Research report series
(University of Michigan. Institute for Social Research.
QC787.P7S867 1988 539.7'3 88-8244
ISBN 0-87944-324-3

ISR Code Number 9028

Published in 1988 by:
Institute for Social Research,
The University of Michigan, Ann Arbor, Michigan

6 5 4 3 2 1

Manufactured in the United States of America

PREFACE

Among the current large-scale federal construction projects planned for the United States, the Superconducting Super Collider (SSC) is financially one of the most expensive, technically one of the most difficult to understand, spatially one of the largest, and scientifically, perhaps, the one that holds the most promise. When compared with other federal projects, many of which are expected to have more adverse impacts on than benefits for their host communities, the SSC project seems relatively benign and even has many positive virtues. And yet, the professional literature on social impact assessment documents that society must be vigilant in assessing the potential social and cultural impacts of predominantly "good" as well as predominantly "harmful" projects.

The most important reason that a social assessment is required for a project like the SSC is the need to determine whether or not there is host community support. There are three reasons why this is necessary. First, the State of Michigan is concerned about its citizens and would not actively seek a potentially harmful project. Second, the United State Department of Energy (DOE) has defined community support as one of its siting criteria. Third, the SSC is not the type of project that would operate effectively in a hostile social environment.

A couple of examples will serve to place this issue in perspective. In 1966, during the siting process of the last great collider--the Fermi National Accelerator Laboratory--the State of Illinois was selected by the Atomic Energy Commission as one of six finalists. Despite the project's recognized benefits, citizens of South Barrington, Illinois--the preferred site--were so completely opposed to the project that Governor Kerner had to withdraw the site from consideration (Lowi and Ginsberg 1976:96). Siting history repeated itself when, in January 1988, local opposition caused Governor Mario Cuomo to withdraw State of New York's proposed SSC site just as the DOE was to announce that the site had made the Best Qualified List.

Host community support is not a static condition, especially for large scale, complex projects that are proposed for rural areas. In such situations, local people begin with few analogs against which to assess the project's potential impacts, and their knowledge of the project increases as more information becomes available. Social assessment research establishes an opinion baseline and begins to systematically document the public and private concerns of local community members in a systematic fashion. When these data are combined with an ongoing program of problem mitigation, many concerns can be addressed so that the local community's evaluation of the project can be based on real rather than perceived impacts. The soundest assessment of community support is produced by an interactive and dynamic process of scientific documentation of potential impact issues and responses with legitimate mitigation commitments. There is evidence to suggest that eventual support for a technically complex project like the SSC derives as much from trust in the mitigation process as from knowledge of the project's impacts.

This report summarizes the results of a study of impact issues, State of Michigan mitigation commitments, and community support for siting the SSC in Michigan. The report reflects 1987 and 1988 research in the Stockbridge area, so there has been an opportunity to measure changes in concerns and support. The scientific findings have stimulated the State of Michigan to seek solutions to potential adverse impacts. For many issues, solutions have been found and consequently attitudes have become more positive towards the proposed project. Because both the scientific research and the mitigation solutions have occurred over the same time period, they have interacted. Thus the report is not only an assessment of potential impacts and a measure of community support, it has become a part of the process by which community support has changed.

Three years of social assessment research findings are contained in this and the first social assessment report (Stoffle, Traugott, Jensen, Copeland 1987), which was also published by the Institute for Social Research. These reports reflect the efforts of a great variety of people all making different and essential contributions. The following is an effort to recognize certain people and to convey the researchers' thanks to all the people who helped.

The authors would like to thank the hundreds of people from the State of Michigan, and Dundee and Stockbridge study areas, who responded to telephone survey and ethnographic questions in the privacy of their homes. The quality of these responses depended on a commitment to complete confidentiality, so the names of these people cannot be used here. More than a dozen township officials served as experts to help define the types of people who most likely would be affected and to select respondents in each of these categories. Edward Grobe, Director of the Ingham County Department of Development, and Wendall Mason, Executive Director of the Jackson Alliance, have been closely associated with the research, sharing maps, lists of key names, and providing essential local perspectives. These two developers have served as "gate keepers," attempting to communicate the benefits of the project and the needs of the people.

The State of Michigan's effort to attract the SSC is governed by the Superconducting Super Collider Commission, chaired by Dr. David Adamany, President of Wayne State University. The Commission funded this research and provided direction through Dr. John Hanieski, Michigan Department of Commerce, and Professor John Mogk, Michigan Energy Resources Research Association and Wayne State University. Dr. Curtis Wiley, of the Michigan Department of Transportation and SSC Project Coordinator for Governor Blanchard, also assisted with the SSC effort. Specialized help was provided by Susan Hohl, Michigan Department of Transportation, Jim Heinzman, Michigan Department of Natural Resources; Jack "Wordsmith" Burdock, SSC office; and Michael Grover and Dan Patrell, Michigan Department of Transportation.

In the private sector, closest ties were with Gilbert/Commonwealth Inc. The project leader there, Steven Ott, worked closely with this research team and provided open access to technical data on all aspects of the project.

In order to conduct the research, the study team needed to understand many essential elements of the SSC proposal and needed to assemble answers to questions raised during field interviews. Technical information was provided by physicists, including Dr. Larry Jones and Dr. Gordon Kane, the University of Michigan; Dr. Bernard Pope, Michigan State University; and several physicists from Fermi National Accelerator Laboratory. Numerous drafts of this report were reviewed for accuracy by Dr. Jones, a feat of some proportion given his residence during this period in California and Europe.

The University of Michigan has supported the research out of the office of Dr. Linda Wilson, Vice President for Research. Special assistance has come from Howard Gobstein. In the Institute for Social Research, financial and administrative assistance was provided by Dr. Raburn Howland, Assistant Director of the Center for Political Studies and Dr. Howard Schuman, Director of the Survey Research Center. Production of the report was made possible by Linda Stafford and JoAnne Lehman at the ISR Publishing Division. Secretarial assistance was provided by Judy Soth and Susan Clemmer.

This preface to the report is being closed with the written thoughts of Orla Sheathelm, a life-long Stockbridge resident, who owns property that would be purchased by the SSC project. Mr. Sheathelm not only speaks for himself, but his thoughts seem to best represent the commonly recorded bittersweet human response to changes brought on by new technologies. The text is from a letter he wrote to be presented at the February 1988, Department of Energy public scoping hearing in Stockbridge and therefore is a public document.

Ladies and Gentlemen,

I'd like to offer my thoughts and observations concerning the impact of the SSC on our community, weighted against the value of the SSC to our nation, and its world position in the fields of science, industry, agriculture and medicine for the remainder of this century and well into the 21st. The environmental impact, as I understand it, would be minor.

On the other hand, the impact economically and socially would be tremendous. Our sleepy little communities would undergo dramatic changes. But, when one considers the changes many of us have seen in our lifetimes, I know that we can contend with the challenges and benefits this project would bring.

I was born early in this century to a hard-working father and mother. We lived in a small frame house, not too well insulated. Many times I woke up in winter to find that snow had blown through cracks around windows and had formed small drifts in the room. We kids would huddle around the stove until Mom had thawed the pump so she could get water for cooking and washing. I walked one and a half miles to country school - rain or shine, heat or cold. It was the greatest education in the world. Farming was hard work, all done with horses and brute strength.

Yes, I can look back on tremendous progress and changes during my life. But they did not occur without being subject to ridicule and skepticism. Many people considered the first car unpractical - it scared the horses, etc. Electricity was taboo in many homes, it would cause many deaths and "Dark Sickness." My dad bought the first rubber tires for farm tractors. The neighbors called him crazy - he'd get stuck all the time and never get his work done. He bought the first combine in our community - another "crazy" thing. Believe it - these changes came about because of Man and his desire for a better life. God knows we wouldn't go back.

My world has seen tremendous progress, and we haven't seen anything yet. I support siting the SSC here in Ingham and Jackson counties. Sure, we will see changes, but as I look back on all those I've seen in my lifetime, I am certain we can meet the new challenges and any changes we see will be for the good as we enter the 21st century.

CONTENTS

APPENDICES

LIST OF TABLES

LIST OF FIGURES

CHAPTER ONE

Executive Summary

1.1 Introduction

At the request of the Governor of Michigan, researchers from the Institute for Social Research (ISR) at The University of Michigan conducted studies of the social effects of and community support for the proposed Superconducting Super Collider (SSC) project in Michigan. Their initial work in 1986 focused on Dundee, in southeastern Michigan, the first site considered by Michigan as a location for the SSC. The State eventually presented proposals for two Michigan sites: the Dundee location and a location near Stockbridge, which is situated in south central Michigan (see Figure 1-1). Research was conducted at both sites.

Because the Stockbridge site was selected in January 1988 as one of the final sites to be considered by the United States Department of Energy (DOE), the majority of this document discusses research conducted at the Stockbridge site. A brief history of the ISR research from 1986 to 1988 is presented in Chapter Two, followed by a detailed discussion of the Stockbridge work in Chapters Three and Four.

The research conducted by the ISR study team was designed to fulfill several needs:

1) to collect scientific data from systematic and in-depth interviews with Michigan residents about their support for, and concerns about, the SSC project;

2) to provide a mechanism for communicating their concerns, in particular to the Michigan SSC Commission, so that the

FIGURE 1-1
MICHIGAN SITES FOR THE SSC
DUNDEE AND STOCKBRIDGE

Gilbert/Commonwealth Inc. of Michigan

State might develop appropriate programs to mitigate these concerns;

and, now that the Stockbridge site (see Figure 1-2) has been selected as one of the seven Best Qualified Sites,

3) to provide input to the United States Department of Energy (DOE) in its scoping activity, as the parameters for evaluating the environmental impacts of the SSC at this site are developed.

The ISR research team defined the concept of "community support" to include two populations: all Michigan residents and residents of the Stockbridge area. This broadened definition of "community" was used to include state residents in acknowledgement of the fact that all Michigan residents would host the SSC. The research program was designed to obtain various types of data from these different populations, in relation to the expected degree of the project's impact on them. The overall design of this data collection effort, now entering its third year, was guided by an extended review of literature on the social impact assessment of other state and federal projects involving high technology (Stoffle, Traugott, Jensen, and Copeland 1987).

As presented in the Stockbridge site proposal submitted by the State of Michigan to the DOE, information was collected from a telephone survey with a representative sample of Michigan residents residing in households with telephones, as well as from an equivalent sample of residents in Ingham and Jackson counties. The questionnaire was designed on the basis of prior work involving in-depth ethnographic interviews that had been conducted in the Stockbridge area.

To summarize the results of the 1987 studies, which are described in detail in the original proposal:

1) Most residents, both in the State of Michigan and the Stockbridge area, support the construction of the SSC and would be amenable to serving as hosts for it;

2) The values to the local economy and the economic base of the State of Michigan were clearly recognized, but a number of concerns and potential adverse effects were also identified.

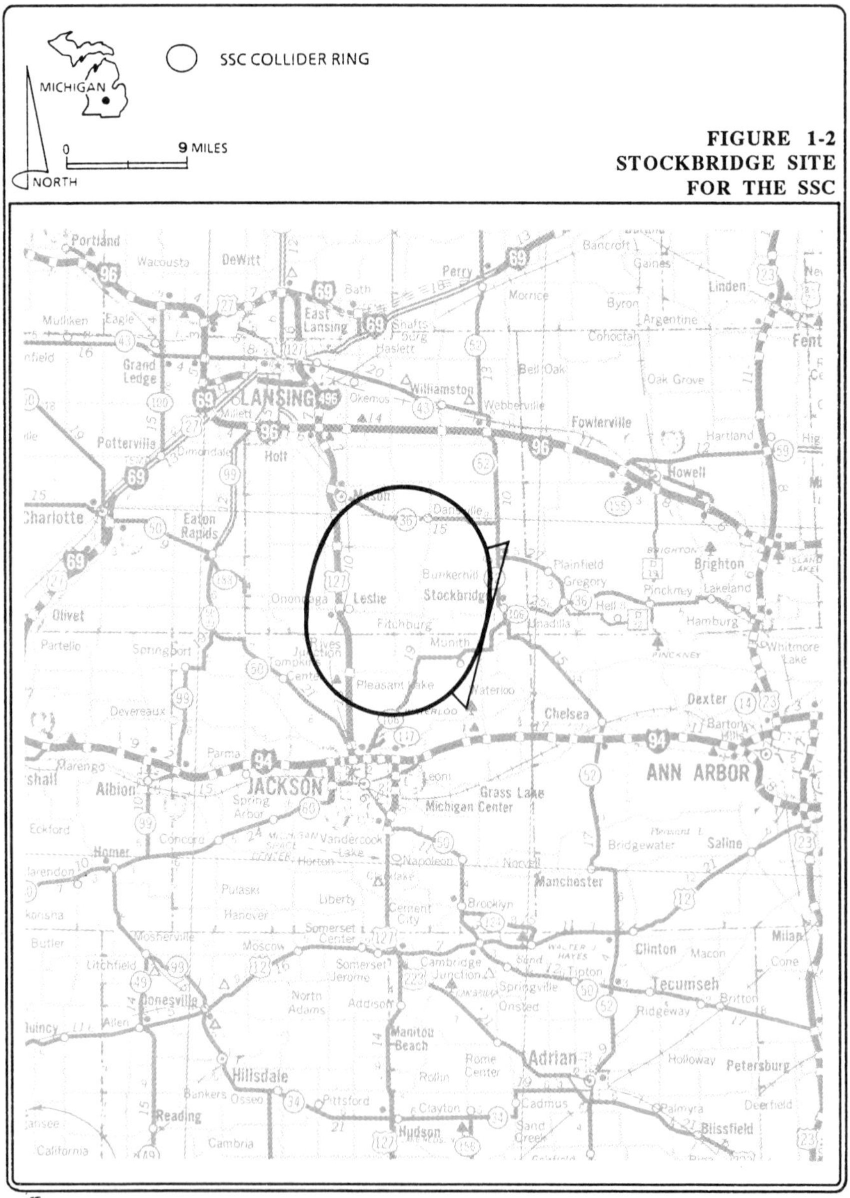

FIGURE 1-2
STOCKBRIDGE SITE
FOR THE SSC

Gilbert/Commonwealth Inc. of Michigan

In response to almost every one of these concerns, the State of Michigan has been able to develop programs and policies to resolve or mitigate the problems raised. This process of interaction between residents, state and local officials, and the Michigan SSC Commission has been a special feature of the proposal preparation process in the State of Michigan.

Since January 1988 when the DOE announced that the Stockbridge site was selected for the Best Qualified List, ISR researchers have started a new phase of research activity. The first component involved recontacting residents in the State of Michigan and the Stockbridge area to ascertain how the announcement and the more real prospect of siting the SSC near Stockbridge has affected their attitudes and opinions. This research effort involved both systematic panel studies of the telephone survey samples as well as in-depth ethnographic reinterviews with representative samples of Stockbridge area residents. Because of this panel design, direct measures of attitude change were possible, as requested by the Department of Energy (Salgado 1988). These studies were undertaken in February 1988.

The second component of the research design included a telephone survey and ethnographic interviews with fresh groups of local residents. The ethnographic interviews with a new group of people were completed in February 1988.

The public opinion telephone survey and the social impact assessment methodologies were designed to provide, in a timely manner, scientifically valid findings about the social impacts of and community responses to the SSC. The data feedback and policy decision points that have influenced the design of the methodology are:

Event	*Reporting Date*
DOE Site Visit	February 16, 1988
DOE Environmental Consultants Scoping Visit	March 14-18, 1988
State of Michigan Contracted for Land Acquisition and Relocation Program	March 8, 1988
DOE Site Visit	May 31-June 3, 1988

Data gathering and analysis steps have been designed so that before a data feedback or policy decision point occurs, some preliminary but valid findings can be presented to the appropriate audience.

1.2 Research on Community Support

The findings summarized below are derived from a comparison of the 1987 and 1988 social impact assessment studies, as well as from ethnographic interviews conducted with a new group of local residents. A draft social impact assessment report was prepared by March 15, 1988 and submitted to the DOE's environmental consultants to be used as part of their data base for the DOE Environmental Impact Assessment. The draft social impact assessment report will be reviewed by state and local people and be finalized by May 31, 1988.

1.2.1 Public Opinion - Telephone Survey

In early 1988, ISR researchers reinterviewed people in order to measure shifts in public support for constructing the SSC in the Stockbridge area. An attempt was made to recontact each of the 605 Ingham and Jackson County residents who responded in 1987, and the 601 state residents who responded last year. Reinterviews were completed with 430 of the local respondents and 349 of the state sample. The design of the 1988 surveys allowed for as many as seven calls to each telephone number from the first survey in an attempt to locate and interview the same person who was interviewed previously.

For each sample the demographic characteristics of the respondents in the 1988 survey were compared with the same characteristics of the 1987 sample. In the Ingham and Jackson County sample, there were no significant differences by age, race, sex, education and income in the two samples. In the state sample, there were differences; therefore, post stratification weights were calculated and used in analysis.

In each case, these reinterviews document a positive shift in opinion toward the SSC. In 1987, 62 percent of the respondents in the local sample favored constructing the SSC in the Stockbridge area, and in 1988, 72 percent favored the project. In 1987, 23 percent of these respondents opposed constructing the Superconducting Super Collider in the Stockbridge area, and in 1988, 11 percent opposed the project (see Figure 1-3).

FIGURE 1-3
PUBLIC OPINION SURVEY
SUPPORT FOR CONSTRUCTION OF THE SSC
IN THE STOCKBRIDGE AREA
LOCAL RESPONSES IN 1987 AND 1988

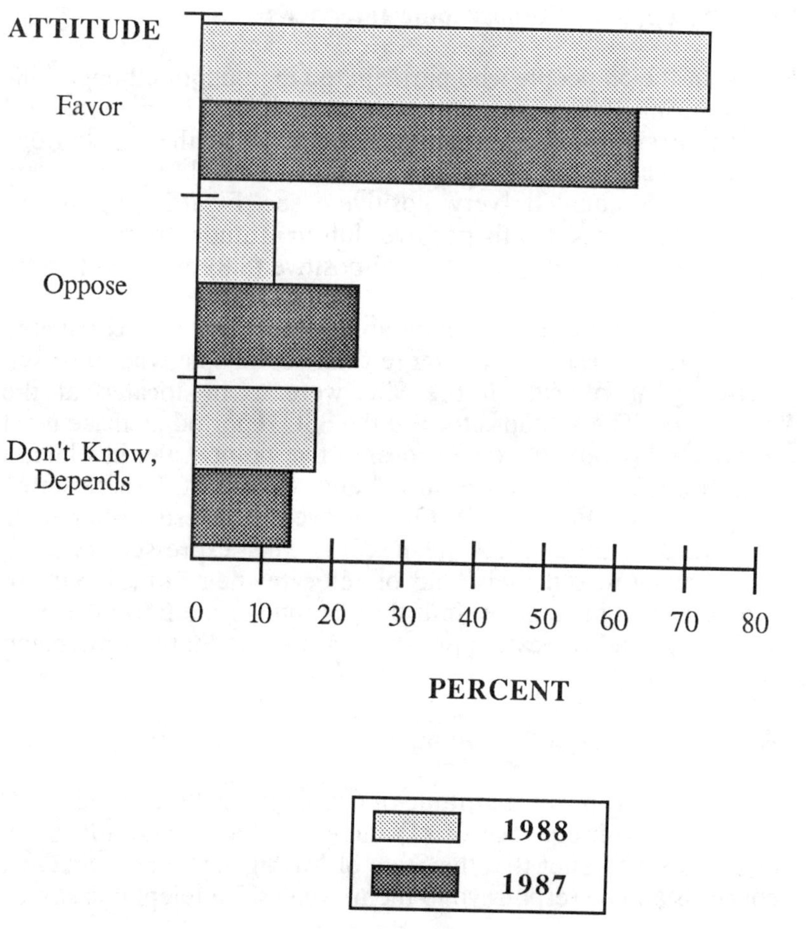

In 1987, 45 percent of the respondents in the local sample said they would be concerned about construction of the SSC in the Stockbridge area, and in 1988, 33 percent said they would be concerned. In 1987, 45 percent of the respondents said they would not be concerned about construction of the SSC in the Stockbridge area, and in 1988, 57 percent said they would not be concerned (see Figure 1-4).

In 1987, 49 percent of the respondents in the state sample said they would be concerned if the SSC were to be built near their homes, and in 1988, 41 percent said they would be concerned. In 1987, 39 percent said they would not be concerned, and in 1988, 47 percent said they would not be concerned (Figure 1-5).

1.2.2 Public Support - Ethnographic Interviews

Twelve of the 55 people who participated in in-depth ethnographic interviews during 1987 were reinterviewed in 1988. The respondents' overall attitudes toward constructing the project in the Stockbridge area were placed on a five point scale in which 1 equaled "very negative" and 5 equaled "very positive" (see Figure 1-6). These reinterviews indicate a slightly positive shift in attitude, from a neutral (2.8) mean attitude in 1987 to a neutral-positive (3.8) mean attitude in 1988.

In 1988, a new group of ethnographic interviews was conducted with 57 people, representing the more than 700 people who must sell their property or relocate if the SSC were to be located at the Stockbridge site. The attitudes toward the SSC reported by these most directly affected people were placed on a five point scale, in which a score of 1 represented "very negative" and a score of 5 represented "very positive" (see Figure 1-7). On the average, these respondents reported a neutral attitude (3.4 mean score). Most expressed concerns related to having to sell their land or relocate their homes, but are withholding a final judgement until they can judge the fairness of the land acquisition and relocation procedures that the State of Michigan will adopt.

1.2.3 Analysis Of Research Findings

The increasingly positive attitude of people in the Stockbridge area appears to be a consequence of (1) increased knowledge about the project and (2) the belief that the State of Michigan will actively elicit their concerns and be responsive to them. Analysis of telephone survey

FIGURE 1-4
PUBLIC OPINION SURVEY
CONCERN ABOUT CONSTRUCTION
OF THE SSC NEAR RESPONDENT'S HOME
LOCAL RESPONSES IN 1987 AND 1988

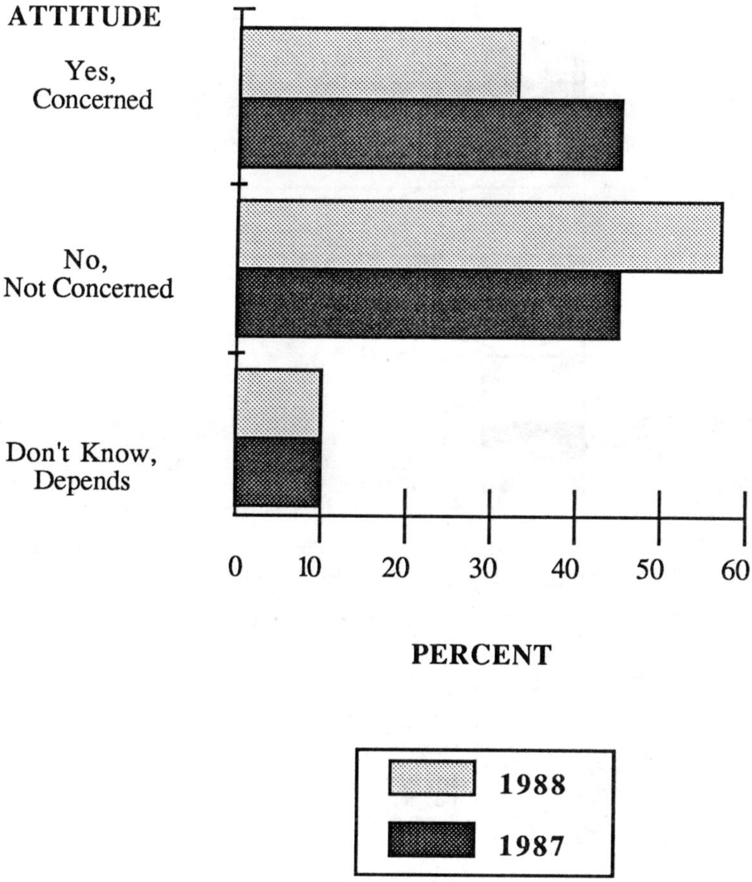

FIGURE 1-5
PUBLIC OPINION SURVEY
CONCERN ABOUT CONSTRUCTION OF
THE SSC NEAR RESPONDENT'S HOME
STATE RESPONSES IN 1987 AND 1988

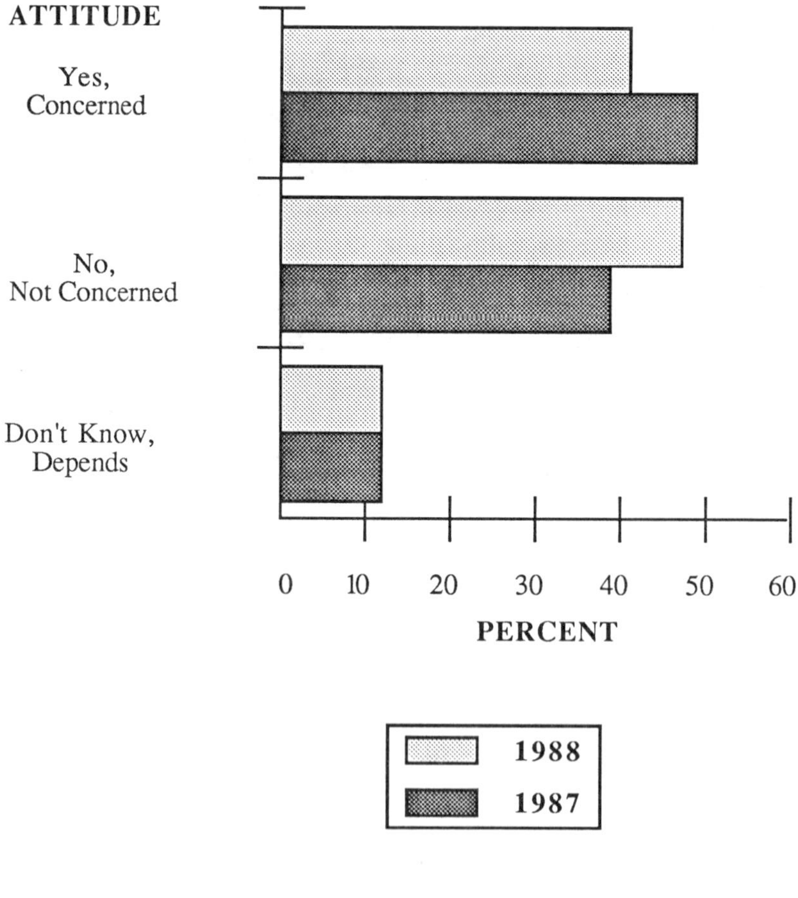

FIGURE 1-6
ETHNOGRAPHIC STUDY
COMPARISION OF ATTITUDES OF PEOPLE ON THE SSC RING TOWARD THE SSC IN 1987 AND 1988

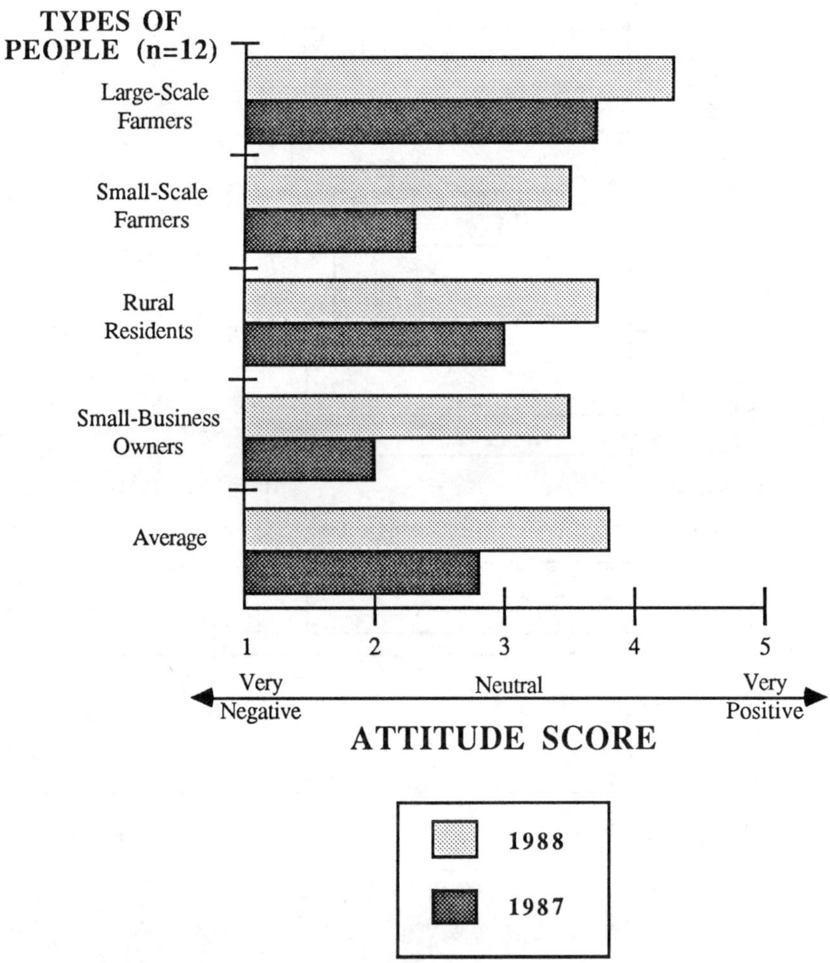

FIGURE 1-7
ETHNOGRAPHIC STUDY
1988 ATTITUDES TOWARD THE SSC BY TYPE OF PEOPLE WHO LIVE ON THE SSC RING

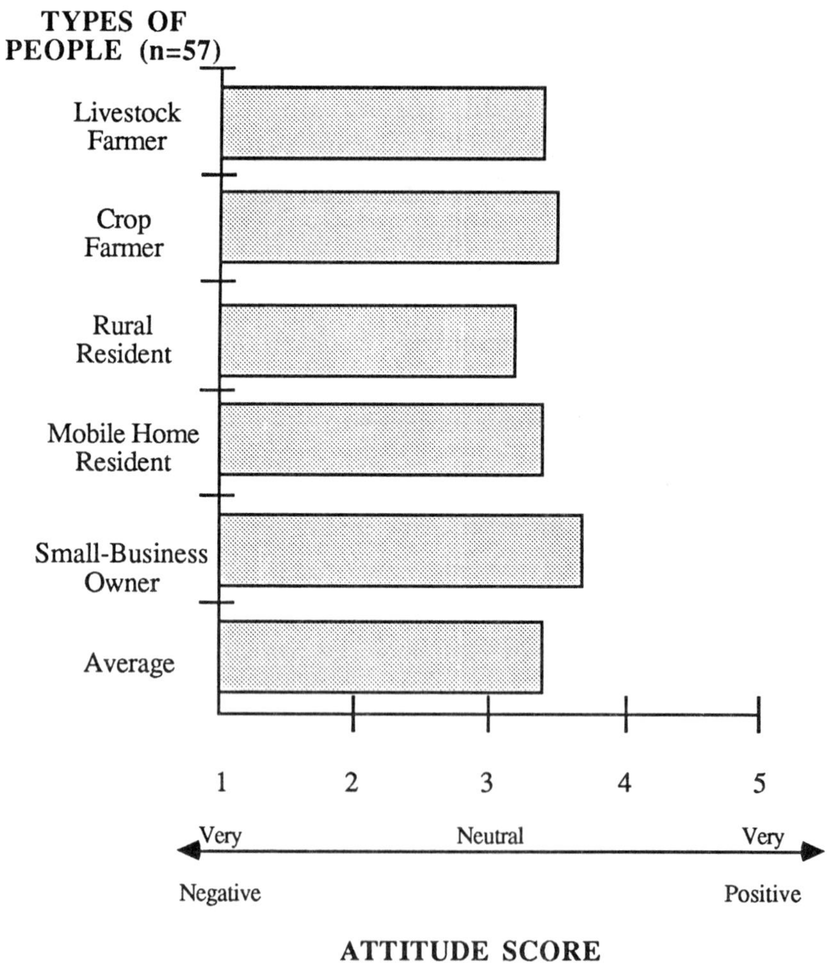

responses demonstrates a direct association between being more knowledgeable about the SSC project and being more supportive of it. Analysis of the ethnographic in-depth interviews suggests that local people are more likely to support the SSC if the State of Michigan respects local sovereignty, is willing to consider the costs as well as the benefits of the project, and mitigates important potential adverse impacts.

1.2.4 State's Responses To Respondents' Concerns

Several concerns about the State's land acquisition and relocation programs were raised during ethnographic interviews conducted in the Stockbridge area. Many of these concerns were raised while the State was in the process of creating its land acquisition and relocation programs; consequently, many have since been addressed. The State was informed of additional concerns and has made a commitment to address them. This information is presented in Chapter Five.

CHAPTER TWO

Project Background

2.1 Introduction

This chapter is divided into two sections that provide background information on the Superconducting Super Collider project and on the 1986 and 1987 phases of research conducted by the ISR team. The information on the SSC is intended to enable readers unfamiliar with the project to better understand the issues raised by respondents during ISR research. The information on ISR researchers' previous work provides a brief history of their involvement with the SSC project in Michigan. The latest phase of research, conducted in 1988, is discussed in greater detail in Chapters Three and Four.

2.2 The SSC

This section opens with a description of the SSC and its purpose. Following is a description of the design process and various phases of the SSC: site selection, construction in Michigan, operation and decommissioning. The section concludes with a discussion of radiation and the SSC.

2.2.1 Description Of The SSC

The Superconducting Super Collider will be the largest scientific instrument ever developed and will enable scientists to study the basic particles from which all matter is formed. Using superconducting magnets, the SSC will accelerate two beams of particles to nearly the speed of light, rotating them in opposite directions around an underground tunnel. The two beams will be made to collide at certain points around the tunnel, where scientists will study the spray of particle components resulting from the collisions. By allowing scientists to peer more deeply into the subatomic world than ever before, the SSC will advance our knowledge about matter in the universe (Universities Research Association nd; Jackson et al., 1986:66-77).

High-energy physics accelerators are used purely for research and are not involved in weapons research or production. In fact, the SSC will be an open laboratory, much like a university. No security clearance will be required of scientists, college students or any others working at the accelerator, and scientists from around the world will be invited there to conduct research. In addition, the public will be welcome to visit the site (Universities Research Association nd).

2.2.2 Design Of The SSC

The United States Department of Energy initiated preliminary Research and Development (R & D) studies for the Super Collider in 1983. The following spring, a Central Design Group was formed to lead the R & D effort and was located at Lawrence Berkeley Laboratory (Universities Research Association 1985:2). During the last few years, the Central Design Group has made several design adjustments, including a change in the length of the tunnel from 52 to 53 miles. At this point, the design is stable, and no basic changes are expected as a result of recent advancements in superconducting technology, (SSC Central Design Group 1986), including advancements that support superconductivity at higher temperatures (*Wall Street Journal* March 23, 1988). The expenditure for Research and Development has been $274 million in 1988 dollars (*Ann Arbor News* March 15, 1988:B2).

2.2.3 Phases Of The SSC

This section discusses four phases of the SSC: Site Selection, Construction of the SSC in Michigan, Operation and Decommissioning.

Phase One: Site Selection

Several states are competing for selection as the site of the SSC. This process began in April 1987, when the DOE issued an Invitation for Site Proposals. This document spelled out the technical criteria to be used in selecting a final site, including geology and tunneling, regional resources, environment, setting, regional conditions and utilities (U.S. Department of Energy 1987a:28-29).

An initial 43 proposals were received and screened by the DOE, which then forwarded 36 of the proposals to the National Academy of Sciences and National Academy of Engineering for their evaluation. In

Sciences and National Academy of Engineering for their evaluation. In December 1987, the NAS and NAE recommended a list of "best qualified" proposals to the DOE. The Best Qualified List included eight states, but New York later withdrew its proposal from consideration. As a result, the following seven sites currently are being considered: Arizona, Colorado, Illinois, Michigan, North Carolina, Tennessee and Texas (see Figure 2-1). The Stockbridge site in Michigan is situated between the cities of Lansing, Jackson and Ann Arbor (see Figure 1-2).

The DOE held Scoping meetings at the seven sites in February 1988, at which the public was invited to express any concerns about the SSC. The scoping meeting for the Stockbridge site was held February 16, 1988. The DOE expects to designate a preferred site in late 1988 and, after an environmental review process, select the final site in early 1989 for recommendation to the president of the United States (U.S. Department of Energy nd).

Phase Two: Construction In Michigan

Phase Two has two elements--Land Acquisition and Construction of the Facility--both of which are described below.

Land Acquisition

The Land Acquisition section covers the land requirements for the SSC, the State of Michigan's program for land acquisition, and the time schedule for acquiring land.

Land requirements. Wherever it is built, the collider will require about 16,000 acres of land. This includes about 8,000 acres for the campus, injector and experiment areas, and 8,000 acres for the stratified fee estate along the tunnel (U.S. Department of Energy 1987a:2,49). The DOE stipulated that land be provided by the states as a requirement for bidding for the SSC. A detailed layout of the SSC at the Stockbridge site is presented in Figure 2-2.

The State of Michigan's land acquisition program. Because of the large land requirements of the SSC, approximately 216 homes and farms will have to be purchased by the State, and approximately 687 property owners will be required to sell their land in the Stockbridge area. (State of Michigan 1987a:17). The State of Michigan is

FIGURE 2-1
BEST QUALIFIED SITES
FOR THE SSC

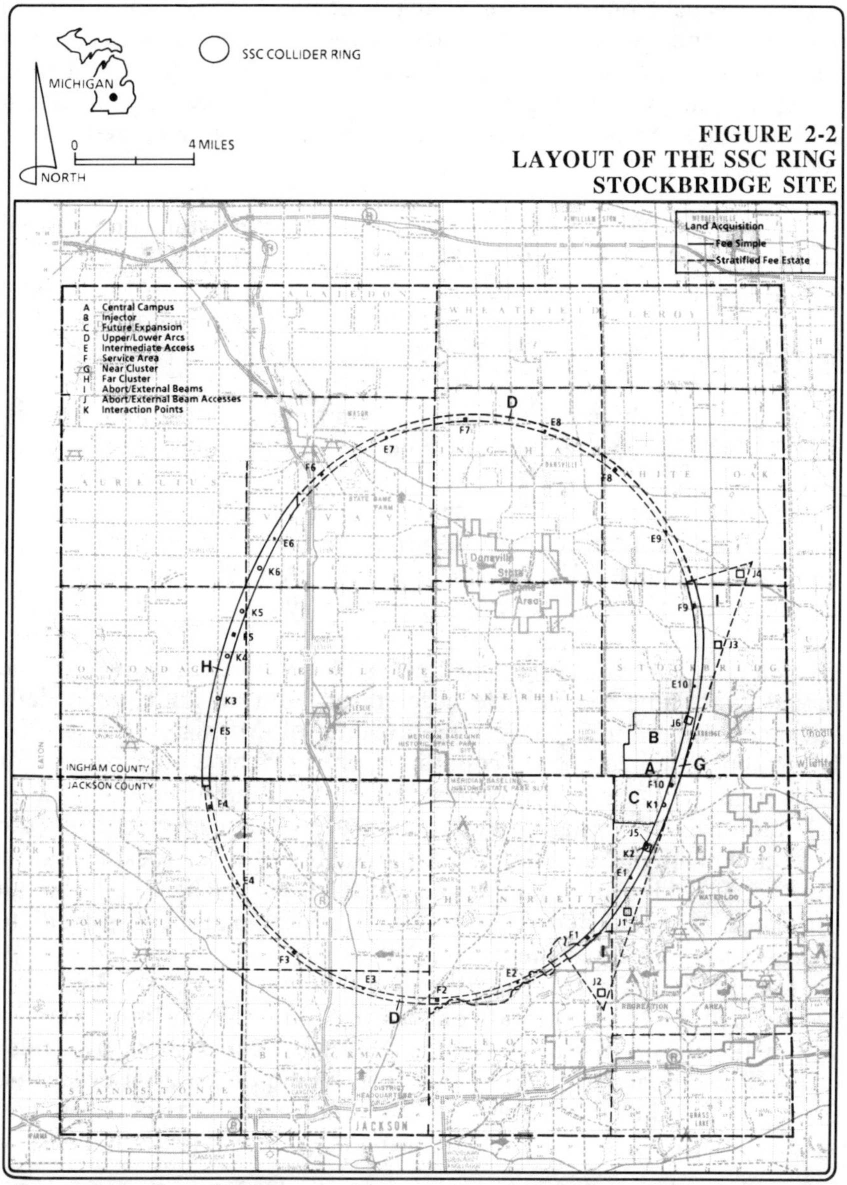

FIGURE 2-2
LAYOUT OF THE SSC RING
STOCKBRIDGE SITE

Gilbert/Commonwealth Inc. of Michigan

developing a land acquisition plan to compensate property owners and residents who would be affected by the SSC (see Appendix F).

The State has a continuing commitment to mitigating adverse effects arising from the SSC siting process. For example, certain parcels of land, owners and/or homes in the stratified fee areas (where only underground access is purchased) may instead require acquisition in fee simple (where surface land is purchased as well). This action could occur if property rights were so impaired that such action was necessary.

Time schedule. Table 2-1 is a schedule for the acquisition of real estate planned by the State of Michigan (Michigan SSC Study Team 1988a).

Construction Of The Facility

This section discusses the construction time schedule and process, and the manpower necessary for completion.

Time schedule. Actual construction of the facility is expected to begin in the summer of 1989. According to the schedule published by the SSC Central Design Group in March 1986, the collider systems should be installed and tested by the mid-1990s.

Process. Tunnel-boring machines will be used to construct the tunnel containing the particle accelerator, with an anticipated daily advance rate of 125 to 150 feet (State of Michigan 1987a:105). Depending upon geological conditions, each machine can proceed for several miles before material must be moved to the surface. A total of 3.3 million cubic yards of excavated material is expected from the project. This figure includes 2.2 million cubic yards of rock, primarily sandstone, and 1.1 million cubic yards of soil. There will be an estimated 25 tunnel shafts around the ring, located at the intermediate access areas and experimental hall sites.

Local quarries have been identified that are used by the Michigan Department of Transportation or private contractors and are generally within 10 miles of the tunnel shafts, with most being about five miles away. These quarries could handle the volume of material excavated from the SSC site. The number of truck hauls necessary for excavation is estimated to be around 68,750, using 60-ton trucks (Birch 1988).

TABLE 2-1
SCHEDULE OF LAND ACQUISITION

SPRING 1988: PRE-APPRAISAL OF PROPERTIES

- Aerial photographis will be taken of proposed areas in Ingham and Jackson counties.

SUMMER 1988: APPRAISAL OF PROPERTIES

- Research will be conducted to determine market value of stratified fee estate property.
- Appraisals will begin.

SUMMER 1988: PRE-ACQUISITION OF PROPERTIES

- Procedures for acquisition and relocation will be developed with special emphasis on properties in the campus, future expansion and service areas and the cluster, as these properties are expected to be the first transferred to the DOE in June 1989.
- A relocation plan will be completed that will be used to determine the individual need of each affected person or family.
- An inventory of available houses and an individual analysis of each displaced person's needs will be completed.
- Meetings will begin to explain the relocation program plus advisory services to assist people who will move.

FALL 1988: NEGOTIATIONS

- Negotiations to obtain options or to purchase properties will begin with owners of properties to be acquired in the first half of 1989.

1989: ACQUISITION BEGINS

- JUNE 1: Titles for land for the first 25 percent of collider properties, including the campus area, will be transferred to DOE.
- Fall: Property Management - Security and maintenance of all purchased properties will be provided. Removal of any acquired improvements will begin.
- September 1: Titles for land for the second 25 percent of collider properties, including the injector area, will be transferred.
- December 1: Titles for land for the third 25 percent of collider properties will be transferred.
- March 1, 1990: Titles for land for all remaining areas will be transferred.

The excavated material can be used for a variety of purposes. These include road beds for SSC access roads, and other purposes of the road commission. Another option would be to have trucks that bring in sand and gravel to make cement for the tunnel cart away the excavated material to fill gravel pits no longer planned for use.

As each sector of the collider tunnel becomes available for use, the technical components will be installed. Pre-operational testing will begin, and the commissioning of each area will proceed sequentially around the ring (SSC Central Design Group 1986:631, 633, 634).

Manpower. The total construction staff for the SSC project includes both on-site workers and personnel at existing high energy physics laboratories. The personnel from other laboratories will assist in the design and construction of such things as the collider magnets (SSC Central Design Group 1986:654).

An estimate of the required manpower is provided in Table 2-2. As mentioned in the table, not all categories of work are included; therefore, these figures are not a total. Instead, they indicate the trend of labor requirements. Management and support and the technical labor requirements for the SSC would peak during the fourth year, while actual labor requirements for construction would peak during the fifth year. Workers in heavy construction, concrete and electrical installation make up the major portion of the construction labor requirement. The construction labor force would fluctuate during the overall construction schedule (SSC Central Design Group 1986:656; Gilbert/Commonwealth Inc. 1986:13,14).

Phase Three: Operation

This section includes a discussion of the appearance of the SSC facilities, employment information and tourism.

Visual Appearance

The SSC's major feature is the collider ring, which, at the Stockbridge site, would be located about 140 feet underground. The collider's oval tunnel will be 53 miles in length and 10 feet in diameter. Figure 2-3 depicts a general view of the SSC ring; however, the layout is reversed at the Stockbridge site. The great length is needed to allow for maximum acceleration of the particle beams. A series of four smaller accelerators will successively speed up the

TABLE 2-2
TOTAL MANPOWER FOR THE SSC PROJECT
(MAN-YEARS)

FISCAL YEAR

	1988	1989	1990	1991	1992	1993	1994
CATEGORIES							
Management and Support	130	281	457	510	390	216	82
EDI (Technical)	379	417	487	499	468	417	200
Construction	48	322	638	1447	1653	1412	135
TOTAL	557	1020	1582	2456	2511	2045	417

Does not include Architectural-Engineering/Construction-Management (AE/CM) and Conventional Systems Construction man-power.

Man-years is defined as one man-year = 2080 man-hours of work.

FIGURE 2-3
A SCHEMATIC VIEW OF THE SSC

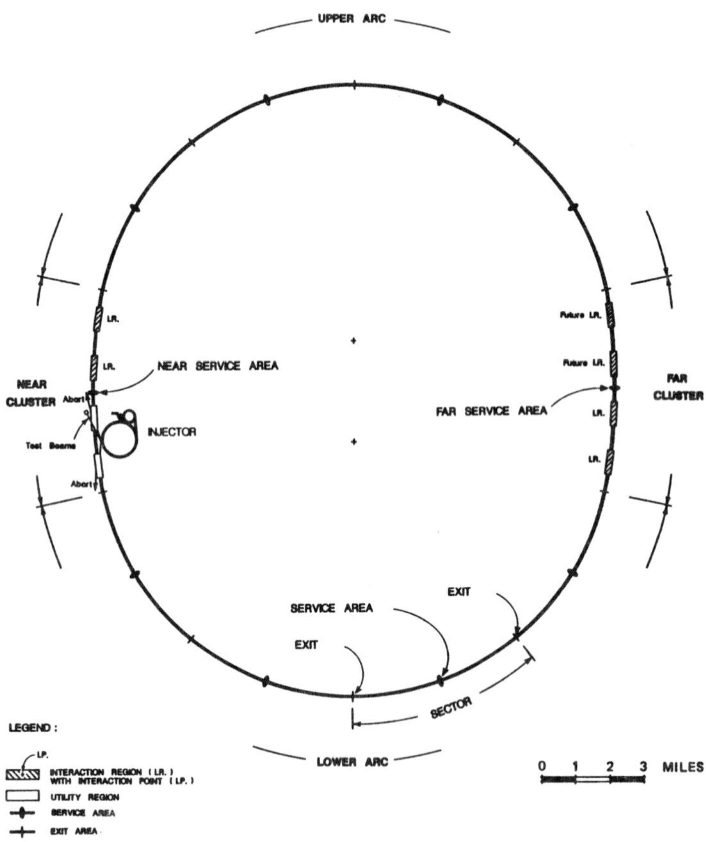

Source: United States Department of Energy, 1987b.

particles to greater and greater energies before injecting them into the collider ring (U.S. Department of Energy 1987a:33).

Because the collider ring will be underground, the enormous scale of the facility will not be readily apparent. The campus area, with its office building, auditorium and various support and industrial buildings, will look like a typical research center or small college campus (see Figure 2-4). Every few miles along the ring, small surface buildings containing such things as support facilities and power supplies will be located (see Figure 2-5). No sign of the system will be seen on the surface of land within the ring, and present activity there will continue (U.S. Department of Energy 1987a:31).

Employment

The SSC will attract a work force of about 3,000 scientists, engineers, technicians and support personnel, including cooks, accountants and librarians. About 2,500 of these people would make up the SSC's permanent work force, and at least 500 would be visiting scientists on short-term stays ranging from a few days to several months (Gilbert/Commonwealth Inc. 1986:14).

However, the number of visiting scientists may be much more than 500. For comparison, about 8,200 people work at the European Center for Nuclear Research (CERN) in Geneva, Switzerland, which is the site of several accelerators. About 3,500 are permanent employees, including scientists, engineers, technicians and support staff, while the majority of the remainder are visiting scientists and students (European Center for Nuclear Research 1987).

Of the 2,500 people that make up the permanent work force, a percentage will be local people who qualify for the types of jobs available at the research facility. This percentage is currently unknown. Fermilab, which is usually considered the closest example of what the SSC will be like, has a program designed to encourage and prepare local people to become employees of the facility. The program involves visits by high school students, internships for college students, and specialized on-the-job training. Although these few observations suggest it is reasonable to expect a similar arrangement for the SSC, systematic research is required before predictions can be made about the numbers of people that would be hired and the types of jobs that would be available at the SSC (Stoffle, Traugott, Jensen and Copeland 1987:56, 59).

FIGURE 2-4
AN ARTIST'S CONCEPTION OF THE FACADE
OF THE CENTRAL LABORATORY BUILDING

Source: United States Department of Energy, 1987a.

FIGURE 2-5
SERVICE AREA SHOWING REFRIGERATOR BUILDING WITH CONNECTING ACCESS SHAFT TO THE COLLIDER RING TUNNEL AND OTHER SUPPORT FACILITIES

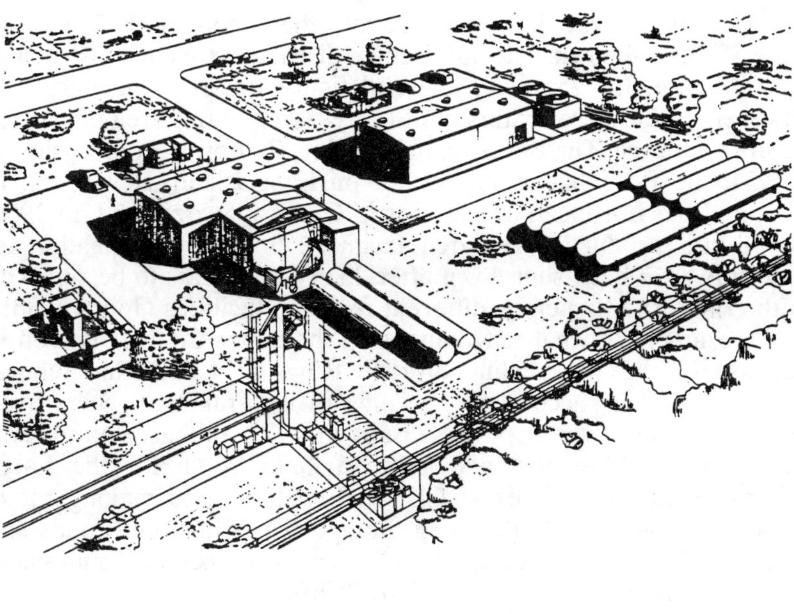

Source: United States Department of Energy, 1987a.

Tourism

One consideration often overlooked regarding the social impacts of the SSC is the fact that the facility will attract tourists to the area. To offer a comparison, about 9,000 people per year take guided tours at Fermilab, while the number of people who take self-guided tours is estimated to be between 20,000 and 35,000 per year. The exact number of self-guided tours is unknown because tourists are not required to sign in when they visit Fermilab (Fermi National Accelerator Laboratory 1988). Tourists at the SSC will include those who visit the area for just a day, and those who visit overnight. The second group would therefore provide business for local hotels, restaurants and other businesses.

Phase Four: Decommissioning

At this time, exact information about the decommissioning phase of the SSC is unknown. The SSC budget discusses costs of construction and operation for 25 years, a useful figure for planning purposes. However, earlier accelerators built for research in this area of science are still operating. The Bevatron in Berkeley, California, commissioned in 1954; the Brookhaven accelerator on Long Island, New York, in 1961; and the European accelerator in Geneva, Switzerland, in 1960, are all still productive research facilities. Some other smaller accelerators have been shut down after 20 or more years to be replaced on the same site by newer, different research facilities (Jones 1988).

An example of what may happen to the Super Collider is seen in the case of the Intersecting Storage Rings at CERN in Geneva, Switzerland. The accelerator is one of several that have been decommissioned over the years. Built in 1971, this accelerator was decommissioned in 1984 and 1985 to make room for the Large Electron-Positron Collider at CERN. In addition to making room, decommissioning the Intersecting Storage Rings liberated funds and manpower, and also enabled the reuse of certain equipment (Radioactivation Task Force 1987:48,69,70).

2.2.4 Radiation And The SSC

Although scientists and government officials have stated that the SSC poses no significant threat due to radiation, questions about radiation safety have been raised by members of the public. Therefore, a brief discussion of radiation and the SSC is included. (For a detailed discussion of radiation and the SSC, see Jones 1986.)

When nuclear particles are accelerated to high energy and then stopped in other material, some of the energy is converted into radiation, and a part of the target is made radioactive. In the case of the SSC, the magnitude of this radiation is equivalent to that produced by a major university with a research hospital and medical school (SSC Central Design Group 1987:ix).

The SSC has been designed to protect the public from exposure to radiation produced by the facility. Two safety zones will accomplish this. First, wherever located, the SSC will be built at least 30 feet underground; and second, the SSC Laboratory will control a 1000-foot-wide zone at the depth of the tunnel. Of that region, 150 feet extending horizontally on either side of the tunnel will serve as shielding. The remaining portion will allow for additional shielding at regions where more penetrating radiation will be produced, as well as for flexibility in the final design and position of the accelerator. This shielding has been designed to keep exposure of the public to no more than 10 millirems per year, well below the legal annual limit of 100 millirems. In addition, a radiation monitoring program will be established (SSC Central Design Group 1987:viii,19,21,25).

The SSC will produce low-level radiation as a result of particle beam collisions, and these beams can be switched on and off like a light. Yet, unlike a nuclear reactor, the SSC will have no inherent source of radiation. There will be no radio-active core of material that could be released in an accident, and there is nothing to sustain a chain reaction. The source of energy will be the same commercial power grid that serves a community's homes (Universities Research Association nd).

2.3 Summary Of Previous ISR SSC Social Impact Assessments

The ISR researchers joined the Michigan SSC Study team in 1986. Their task was to provide a better understanding of the potential community responses to and social impacts of a proposed SSC site in southeastern Michigan, the first of two sites that Michigan eventually proposed as the location for the SSC. The study team wanted a preliminary assessment of the public attitudes toward the SSC at that time, the types of local social impacts that might be expected, and some understanding of the methodologies that would be used to conduct a social impact assessment study (Stoffle and Traugott 1986:2). The following text discusses this previous ISR research.

2.3.1 Literature Review

As an initial step, a review was conducted of past research on public attitudes toward a number of related issues and projects that would be useful in designing surveys if the project were to be pursued. These past studies included assessments of the general public's views toward science and technology, the role of technology in economic development, and a willingness to use public funds to support large-scale projects such as the SSC. The conclusions derived from this literature review were useful in the development of the telephone surveys of public opinion. These conclusions were as follows:

1) A majority of American citizens have a positive predisposition toward science. This is related to education and the amount of attention people pay to science. The better educated and more attentive to science issues, the more favorably inclined they are toward science issues and technology.

2) Compared with 30 years ago, the American public has become more apprehensive about science and technology. Their confidence in scientists has declined, although this is part of a general decline in confidence in a number of American institutions.

3) The American public has difficulty in distinguishing between basic and applied research, and they are less likely to be willing to invest funds in basic research than in areas where practical outcomes are the likely result. This is especially true if the public perceives tradeoffs or alternative programs in which the funds might be invested.

4) There were no studies that related directly to the SSC project because of its many unique components. The literature on public reactions to other scientific issues such as fluoridation, nuclear power plant construction, and other environmentally sensitive projects indicates the basis for opposition to these projects. Important factors include (a) the perception of danger, including both risk and harm, (b) a lack of full understanding of the project, and (c) alienation from society. Supporters of development projects often have been able to deal with the first two of these issues with public information programs, and the effectiveness of such programs can be measured in subsequent surveys. In other instances, the public response to a project was more accurate and informed, and the project was eliminated or modified because of poor project design or inappropriate project impacts on the environment (Stoffle, Traugott, Jensen and Copeland 1987).

2.3.2 Phase One: Dundee Study - 1986

The initial social impact assessment study focused on a single study area near Dundee, Michigan, located southwest of Detroit. The research defined many of the basic variables and the types of people that would be central to understanding the effects of the SSC in southeastern Michigan. In-depth interviews were conducted with 49 farmers, professionals and townspeople. Different social impacts were perceived by farmers, townspeople, men and women. In addition, the study explored various research methodologies that would be most appropriate for further analysis of SSC effects and Michigan citizens' responses to the project (Stoffle, Traugott, Jensen, Copeland 1987).

2.3.3 Phase Two: Dundee And Stockbridge Studies - 1987

The second phase of research began in the spring of 1987 after the President of the United States announced that the SSC was a national scientific priority and would be constructed. The Governor of Michigan announced his commitment to making a bid to locate the SSC in Michigan and recommended consideration of a second site. Thus the State was committed to providing the scientific documentation for two complete siting proposals. A location near the village of Stockbridge was selected as the focus of the second set of site studies.

The DOE issued an Invitation for Site Proposals (ISP) that defined the criteria for the evaluation of the states' siting proposals and the criteria for the development of a short list of "Best Qualified Sites" (DOE 1987). These guidelines specified that each proposal should include an assessment of the public commitment to host the SSC:

> Community support will play an important role in the ultimate success of the SSC. The SSC staff will actively seek the support of surrounding communities by establishing a good-neighbor policy. It is important to determine at an early stage what community support exists and how strong any opposition might be to siting the SSC in a region (United States Department of Energy 1987a:54).

Based in part on this ISP criteria, the State of Michigan requested that the University of Michigan conduct another study in order to provide scientifically valid evidence of community support of and opposition to the SSC facility. In addition, the State was motivated by the desire to know of potential adverse local effects so that they could be addressed before the siting proposal was sent to the DOE.

Telephone Survey

During this second study, ISR researchers conducted telephone interviews to assess public opinion at the local (Dundee and Stockbridge sites) and statewide levels. This work involved surveys of representative samples of residents using a questionnaire developed in the previous study and based upon field data collected from the Dundee site. At the state level, 601 respondents were interviewed, representing a response rate of 63 percent. In Jackson and Ingham counties 605 respondents were interviewed, representing a response rate of 64 percent. In Monroe and Lenawee counties 602 respondents were interviewed, representing a response rate of 63 percent.

Ethnographic Study

In addition to the telephone survey, in-depth interviews were conducted with local people at both sites.

Stockbridge Site

Ingham and Jackson County residents were asked for their reactions to the proposed SSC. Background information about the locale was gathered from interviews with key cultural experts in the area. Since most of the land required for the SSC would be farmland, more responses were sought from farmers than townspeople. These two types of landowners were further classified on the basis of the on-site or off-site location of their property. In all, 55 individual interviews were conducted: 40 with farmers, 15 with townspeople and business people.

Dundee Site

At the Dundee site, discussions with focus groups composed of local government officials and small-business owners were conducted. Each focus group discussed and evaluated the impacts of various scenarios of SSC construction and operation on local governments, small businesses and the entire community. These focus groups consisted of representative types of government and business people. In all, 26 local government and business people participated in focus group meetings.

2.3.4 Phase Three: Stockbridge Study - 1988

Because the Stockbridge site was selected as one of the seven Best Qualified Sites, the ISR research team continued to assess public opinion about and reaction to the project. The following chapters discuss the methodology of this research design and the study findings in detail.

CHAPTER THREE

Public Opinion Telephone Survey Methodology And Findings

3.1 Introduction

This chapter describes the opinions of Michigan residents who participated in surveys about the Superconducting Super Collider. Participants were asked a variety of questions about Michigan seeking and hosting the SSC, and their opinions were tabulated to provide a statistical analysis. To obtain this information, telephone interviews were conducted with a random sample of state residents and a random sample of people residing in the Stockbridge area. The telephone surveys are the first of two sources of information about the potential social impacts of and citizens' responses to the SSC project. The second, and more qualitative, source of information is ethnographic interviews conducted with the people most immediately and directly affected by the project. The ethnographic findings are presented in Chapter Four.

This chapter is organized into three sections. The first section discusses the methods by which people were chosen to participate in the study and how they were interviewed. The second section presents findings about the opinions of the sample of people interviewed from the Stockbridge area. The third section presents findings about the opinions of a statewide sample of Michigan residents. The chapter concludes with a list of key issues raised during these telephone interviews.

3.2 Survey Methodology

The research team assessed citizens' responses to the SSC in the State of Michigan and in the Stockbridge site area (see Figure 3-1) through the use of telephone interviews with representative samples of residents (see Appendix A and Appendix B). The results of these surveys provided an ongoing monitoring of important issues related to the siting of the SSC facility in the Stockbridge area and guidance to the State of Michigan in developing programs to mitigate these concerns.

The design for these surveys involves a panel study in which respondents from each of the two samples interviewed in July 1987 were recontacted in early 1988 to see whether their knowledge of and attitudes toward the project had changed, and in what ways. This is an indirect way of measuring how successful the information campaign and mitigation programs of the State of Michigan have been.

In the work conducted in the summer of 1987, a sample of 605 respondents were interviewed in the Jackson and Ingham counties and 601 interviews were conducted among a statewide sample of Michigan residents. In the two-county sample, telephone households in the city of Lansing, the state capital, were excluded from the sample because of its high urban density and its distance from the site. Information about these respondents obtained at the time of these initial interviews was used as the basis for recontacting them approximately eight months later. Because of the limited time available between the announcement of the Stockbridge site as one of the seven best qualified and the scheduled visit of the Department of Energy team for the scoping session, the recontact interviews were conducted as rapidly as possible and under less than optimal field conditions. That is to say, advance letters were sent to the respondents at their previous known addresses, and recontacts were limited to no more than seven attempts. The field period for each survey was limited to approximately two weeks elapsed time. The recontact interviews with the Stockbridge respondents were conducted between February 1 and 14, 1988; the interviews with respondents in the state sample were conducted between February 22 and March 1, 1988. The questionnaire was modified only slightly from the original, in order to allow for the new status of the Stockbridge site on the Best Qualified List and the increased likelihood that the SSC could be sited in Michigan.

The major methodological problems associated with panel (or recontact) studies are the reliability of the information available about

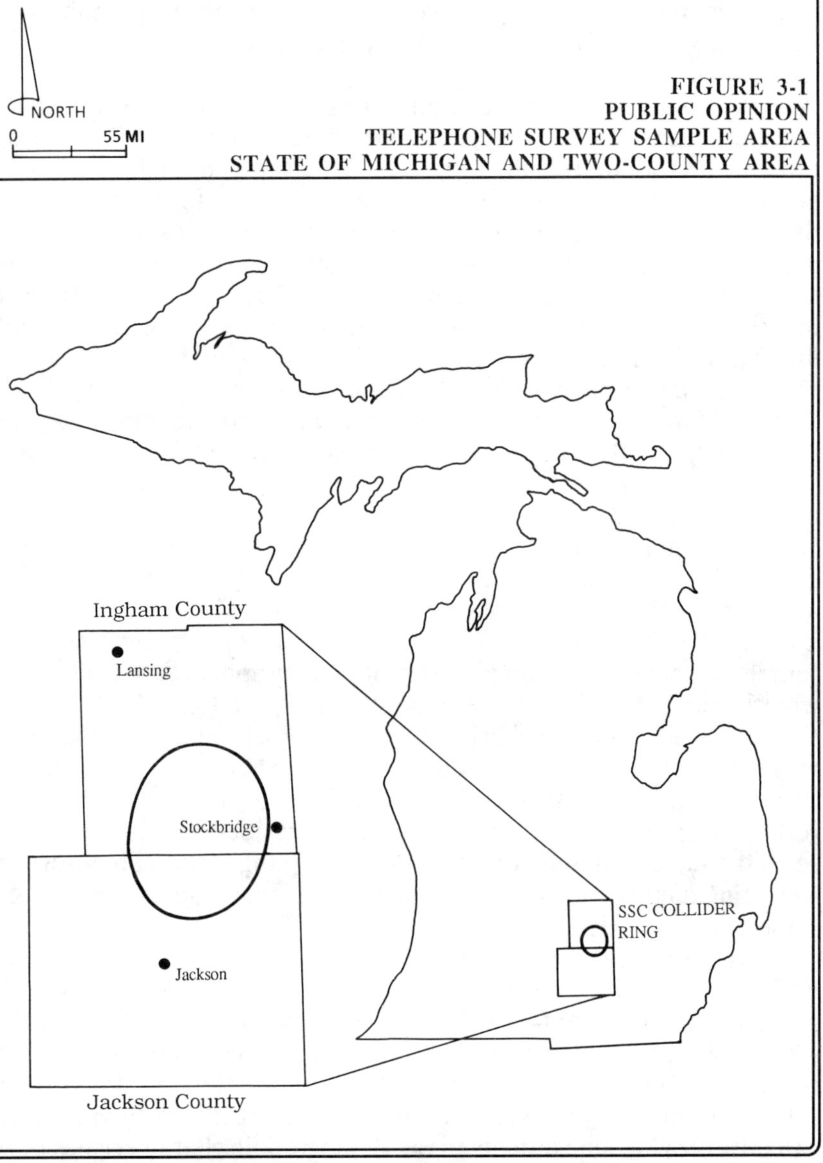

FIGURE 3-1
PUBLIC OPINION
TELEPHONE SURVEY SAMPLE AREA
STATE OF MICHIGAN AND TWO-COUNTY AREA

Gilbert/Commonwealth Inc. of Michigan

the respondent and the high mobility of the United States' population. In order to maximize recontact rates, information should ideally be collected about other individuals who might know the whereabouts of a respondent at a future date. Because of limited time in which to ask questions on the original survey and an unknown probability of conducting the second survey, this information was not collected in the summer of 1987.

The population of the United States is highly mobile, with about one-fifth of households changing their address every year. The rates of mobility are not uniformly distributed across the population, however; people living in larger, urban areas are more likely to move than those living in rural areas. This led to the expectation that under any fixed design, the recontact rate would be higher for the Stockbridge area sample than for the state sample because of its rural nature compared to respondents from highly urban areas of the state as a whole, especially from the city of Detroit. This was in fact the case, as recontact was made with 430 respondents in the original Stockbridge sample (a recontact rate of 71%) compared to recontact with 349 respondents in the original state sample (a recontact rate of 58%). These rates are typical for recontact surveys conducted with the design parameters described above.

The primary significance of these recontact rates has to do with the degree to which the respondents who are interviewed the second time are representative of those who were interviewed initially. The respondents in the initial and recontact surveys in each sample were compared on a variety of demographic variables to see to what extent differences were apparent. In a relatively homogeneous population as in the Stockbridge area, the prospects for comparability are higher; and, in fact, the differences between the distributions of demographic characteristics (age, sex, income, education, length of residence in the community, and place of residence) of the first and second time respondents were within sampling error. Therefore, no weighting adjustment was made for relative nonresponse according to these characteristics.

In the case of the state sample, however, this same analysis showed differences in the distributions of these characteristics that were greater than expected from sampling error alone. As a result, the data were reweighted to bring the demographic characteristics of the recontact sample into line with these parameters in the original sample. While this is a standard procedure in instances such as this, it can only provide adjustments in the demographic characteristics of the respondents. There is no statistical adjustment which is possible based upon attitudes or opinions which the respondents held at the time of their first interview, or for relationships between such variables.

In 19 out of 20 cases using this design, the estimates of percentages derived for the entire sample from either survey would be expected to be within 5 percentage points of the actual percentage in the population. For subsamples, the differences would be larger.

3.3 Local Study Area Findings

Analysis of data from the recontact interviews with the Stockbridge area sample suggest that efforts to inform residents about the SSC project and state programs to mitigate their concerns have been successful. In the original survey conducted in July 1987, it was reported that those who were most informed about the project were the most likely to support the SSC. This fact lead to the recommendation that an extensive educational program be undertaken. Results from the recontact interviews suggest that knowledge levels about the project increased. In addition, support for the project is increasing and concern about construction is decreasing.

In the original survey, about one-third of those interviewed (36 percent) had heard or read about the SSC, and a substantial majority of those surveyed (79 percent) indicated they felt the State should try to get the project. In the latest wave of recontact interviewing, 90 percent have heard or read about the SSC. Almost all of these respondents (80 percent) are aware that Stockbridge is the proposed site in Michigan where the SSC might be built; and in fact, one-third can recall at least one other state with a site on the Best Qualified List.

Data are presented in Table 3-1 that show changes in the level of concern about construction of the SSC near the homes of Stockbridge area residents from July 1987 to February 1988. Among those recontacted, there had been 45 percent who were concerned about construction last summer, and now only 33 percent of those recontacted people are concerned. At the same time, 45 percent had indicated they were not concerned, while now there are 57 percent who express that same opinion.

TABLE 3-1

CHANGES IN CONCERN ABOUT CONSTRUCTION OF THE SSC NEAR THE HOMES OF STOCKBRIDGE RESIDENTS, JULY 1987 TO FEBRUARY 1988

	ATTITUDE IN JULY 1987			
	Yes, Concerned	Don't Know, Depends	No, Not Concerned	Total (1988)
ATTITUDE IN FEB. 1988				
Yes, Concerned	53%	35%	13%	33%
Don't Know, Depends	12	14	8	10
No, Not Concerned	35	51	79	57
	100%	100%	100%	100%
	(n=195)	(n=43)	(n=192)	(N=430)
Total (1987)	45%	10	45	100%

The exact question wording is:

The SSC will be built in the form of a large circular ring about 53 miles in length that is primarily underground. Would you be concerned if it were going to be located near your present home?

A further breakdown of the data presented in Table 3-1 reveals that of those who were concerned in July 1987, only 53 percent still are concerned, while 35 percent are no longer concerned. Of those who were not concerned in July 1987, a full 79 percent still are not concerned, while 13 percent are now concerned.

Data are presented in Table 3-2 that show the relationship between support and opposition to the SSC in July 1987 compared to the results from the February 1988 survey. Among those recontacted, there had been 62 percent that had favored the project in July 1987, and now 72 percent favor the project. In July 1987, 23 percent opposed the project, while in February 1988, 11 percent opposed the project.

A further breakdown of the data presented in Table 3-2 reveals that attitudes among those who favored the project initially remain strong, while there has been some shift toward support among those who opposed the project six months ago. For example, of those who were opposed last summer, 41 percent now favor the project. At the same time, only 3 percent of those who favored the project last summer now oppose it. The preponderance of changes in opinion have been toward support for the SSC, again reflecting the effects of the dissemination of information about the project.

3.4 State Findings

Analysis of data from the recontact interviews with the statewide sample indicate that more state residents are knowledgeable about the SSC. In the initial survey, 22 percent of the state residents interviewed had heard or read about the SSC. In the recontact interviews, 77 percent had heard or read about the project.

In the recontact survey, state residents were asked whether they favor or oppose construction of the SSC in the Stockbridge area. Seventy-two percent of the respondents favored construction, 14 percent opposed, and 14 percent answered "don't know, or depends." This question was not asked in the initial survey; therefore, no comparable data are available.

In both the 1987 and 1988 surveys, state residents were asked if they would be concerned if the SSC were to be located near their home. In this state panel, the shift toward decreasing concern observed for the local sample was also observed, although the magnitude of the decline in level of concern was not as great as among Stockbridge area residents (8 compared to 12 percentage points). In the 1988 survey,

TABLE 3-2

CHANGES IN SUPPORT FOR CONSTRUCTION OF THE SSC IN THE STOCKBRIDGE AREA, JULY 1987 TO FEBRUARY 1988

	ATTITUDE IN JULY 1987			
	Favor	Don't Know, Depends	Oppose	Total (1988)
ATTITUDE IN FEB. 1988				
Favor	85%	65%	41%	72%
Don't Know, Depends	12	22	27	17
Oppose	3	13	32	11
	100%	100%	100%	100%
	(n=266)	(n=62)	(n=102)	(N=430)
Total (1987)	62%	14	23	100%

The exact question wording is:

> Overall, would you favor or oppose the construction of the SSC in the Stockbridge area? Would you strongly or not so strongly (favor/oppose) the construction of the SSC?

For simplicity, the categories of "favor strongly" and "favor, not so strongly" were combined into "favor." The categories of "oppose strongly" and "oppose, not so strongly" were combined into "oppose."

there were fewer concerned respondents than unconcerned, a reversal of the finding from last summer. As the data in Table 3-3 show, 41 percent of the respondents in 1988 were concerned, and 47 percent were unconcerned. In 1987, 49 percent of the respondents were concerned, while 39 percent were unconcerned.

A further breakdown of the data in Table 3-3 indicates that of those concerned in 1987, 65 percent are still concerned, while 27 percent of those who were concerned are no longer concerned. Of those who were not concerned in 1987, 74 percent remain unconcerned, while 17 percent are now concerned.

3.5 Public Opinion Regarding The SSC: Volunteered Responses To Telephone Survey Questions

While support for the SSC facility has been growing in the local area, there remain significant issues about the project that concern local citizens. These were captured in open-ended responses to questions about perceptions of the major benefits and/or disadvantages of having the SSC located in the Stockbridge area. The statistical description of these concerns is presented below. These are offered as indicators of the nature of concerns and as matters that should be considered and responded to as part of the Department of Energy's environmental impact assessment process. The percentages listed below add to more than 100 percent because of multiple mentions.

3.5.1 Benefits To Having The SSC

In response to the question, "What do you think would be the major benefits of having the SSC facility located in the Stockbridge area?," these were some of the replies, along with the frequency of the response for the local and the statewide surveys. The replies are listed in descending order of the local survey's frequencies (see Figure 3-2).

Create More Jobs

Employment levels would go up; it would bring in work for SSC itself; it would add jobs. (Local 56%; State 38%)

TABLE 3-3

CHANGES IN CONCERN ABOUT CONSTRUCTION OF THE SSC AMONG RESIDENTS IN THE STATE OF MICHIGAN, JULY 1987 TO FEBRUARY 1988

	ATTITUDE IN JULY 1987			
	Yes, Concerned	Don't Know, Depends	No, Not Concerned	Total (1988)
ATTITUDE IN FEB. 1988				
Yes, Concerned	65%	24%	17%	41%
Don't Know, Depends	8	27	9	12
No, Not Concerned	27	48	74	47
	100%	100%	100%	100%
	(n=195)	(n=43)	(n=192)	(N=349)
Total (1987)	49%	12	39	100%

The exact question wording is:

> The SSC will be built in the form of a large circular ring about 53 miles in length that is primarily underground. Would you be concerned if it were going to be located near your present home?

FIGURE 3-2
PUBLIC OPINION SURVEY
PERCEIVED BENEFITS OF THE SSC
1988 LOCAL AND STATE RESPONSES

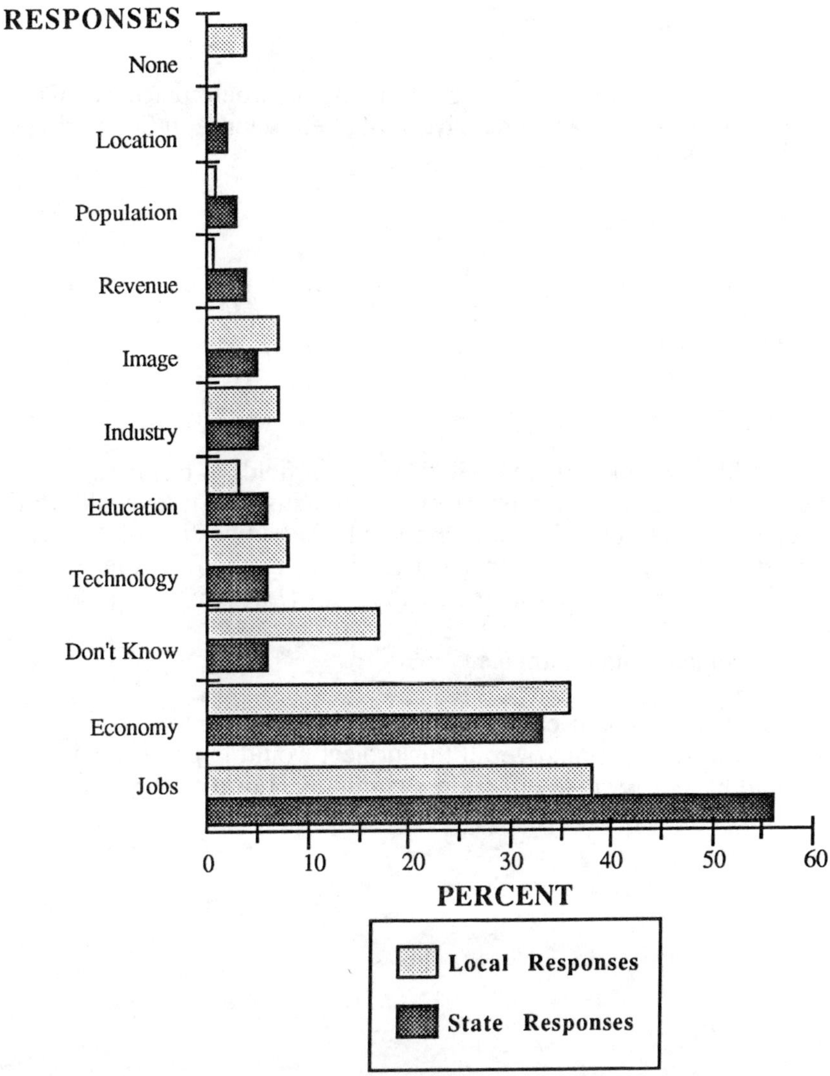

Improve The Economy

It would benefit the economy in the Stockbridge area of Michigan; the economy as a whole would improve. (Local 33%; State 36%)

Don't Know

There were 6 percent of the local respondents and 17 percent of the state respondents who answered "don't know" when asked about the benefits of having the SSC.

New Technology

New technology would be taught in schools; would place Michigan in the forefront of new technology; would help science in Stockbridge. (Local 6%; State 8%)

Educational Value

The SSC would bring more educated people into our community; would get people with good backgrounds; raise standards of education. (Local 6%; State 3%)

Develop A New Industry In This Area

The SSC would bring more business; new fields of business; would change Michigan from a motor state to an industrial state; could draw other supplemental industries that would be supportive of the SSC project or similar projects that would be an offshoot of the SSC; a spin-off effect would be created by the SSC. (Local 5%; State 7%)

Status/Enhance State's Image

The project would enhance the State's image; it would give our State world recognition; overall this project would improve Michigan; it would have a positive effect on everything. (Local 5%; State 7%)

Tax Revenue

The SSC would bring in a lot of revenue; the increased tax base would help local governments. (Local 4%; State Less Than 1%)

Increase The Population

The SSC would bring more people into the area; more people would be living in the community. (Local 3%; State 1%)

Convenient Location

The SSC would be centrally located between the two universities; the SSC would be out of the way, not close to the houses or the main highways; it's an area that could handle the proposed construction without any problems. (Local 2%; State 1%)

None

There were no local respondents and 4 percent of the state respondents who indicated there would be no benefits.

3.5.2 Disadvantages/Major Concerns About The SSC

In response to the question, "Would there be any disadvantages to having the SSC facility located in the Stockbridge area?," these were the most frequently offered replies, along with the frequency of the response for the local and the statewide surveys. The replies are listed in descending order of the local survey's frequencies (see Figure 3-3).

Don't Know

One-third (32 percent) of the local respondents and 38 percent of the state respondents said they did not know of any disadvantages to having the SSC.

None

Eighteen percent of the local respondents and 23 percent of the state respondents said there were no disadvantages to having the SSC.

FIGURE 3-3
PUBLIC OPINION SURVEY
PERCEIVED DISADVANTAGES OF THE SSC
1988 LOCAL AND STATE RESPONSES

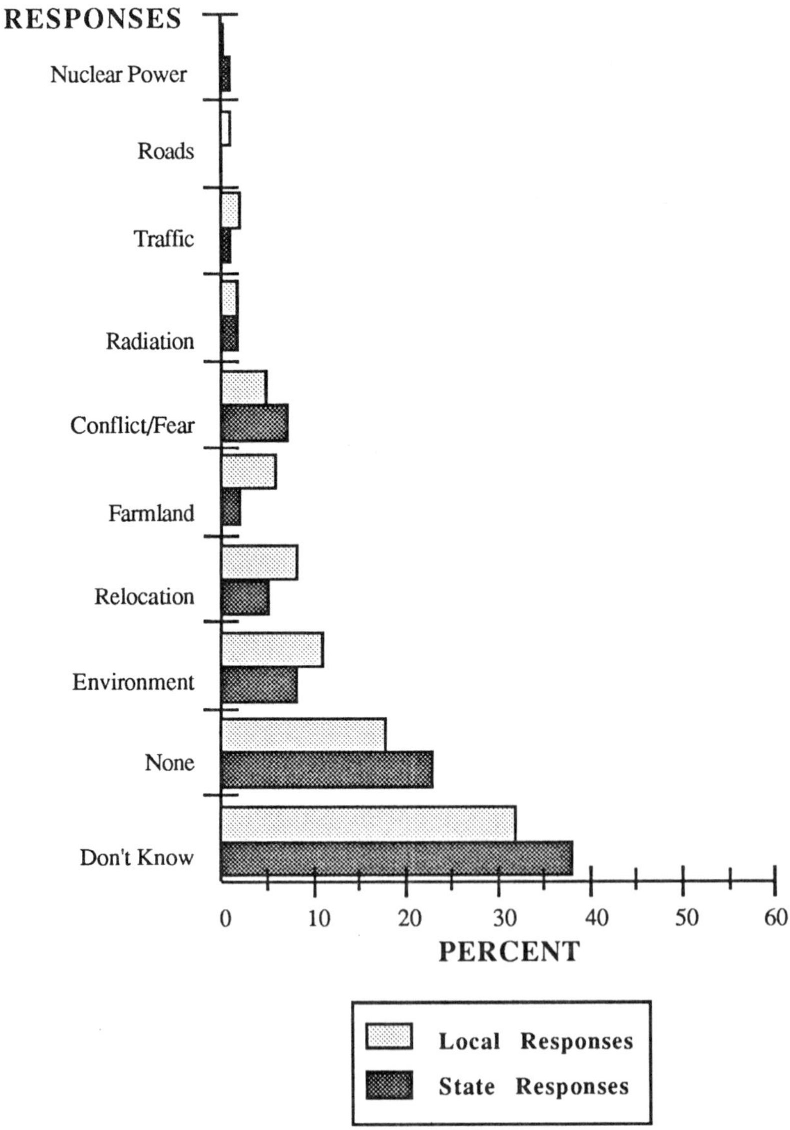

Environmental Harm

The SSC might have some effect on the water supply and perhaps pollute the water and pollute the air. (Local 11%; State 8%)

Relocation Of Home Owners

Relocation would disturb homes; some people would have to move and this could cause some problems for some families. (Local 8%; State 5%)

Loss Of Farmland

The SSC would waste of a lot of farmland; a lot of farmland would be used for the project; the farming community could be in danger; the SSC might put the farmers out of business. The Stockbridge area would change from quiet and rural to high-technology. (Local 6%; State 2%)

Conflict In The Community/Fear Of Technology

The people will be frightened of the technology; there will be conflicts in the community; uncertainty about how the SSC would operate in the future. (Local 5%; State 8%)

Radiation

There could be health risks due to radiation from the SSC. (Local 2%; State 2%)

More Traffic

Traffic would increase because of the SSC and the resulting population growth. (Local 2%; State 1%)

Roadway Construction

The roads may have to be increased and enlarged to accommodate SSC construction vehicles and activity. The roads would become crowded due to construction. (Local 1%; State 0%)

Generation Of Nuclear Power

People are concerned about and afraid of nuclear waste generated by the SSC project. (Local Less than 1%; State 1%)

3.5.3 A Comparison Of 1987 And 1988 Volunteered Responses

A trend analysis of the respondents' answers to an open-ended question about their perceived benefits/advantages, as well as their concerns and perceived disadvantages, to having the SSC located in the Stockbridge area reveals both more moderate expectations of benefits and a decreased level of concern. Both of these phenomena seem appropriate in light of the extensive flow of information which took place between the two interviewing periods and the time that was available for the local citizens to absorb and evaluate it.

On the matter of perceived benefits, more respondents offered some type of response in the second survey compared to the first. The proportion offering no response or indicating that they did not know of any advantage/benefit declined from 17 percent to 6 percent across this period (see Figure 3-4).

The most frequently cited advantage/benefit remains employment opportunities, as 56 percent of the respondents each time indicated they expected there to be more jobs available and employment in the area would go up. Not as many respondents expect a strong boost to the local economy, however. In the 1987 survey, this response was given by 53 percent of those surveyed; in the 1988 reinterview it was offered by only 33 percent. Positive shifts in the size or composition of the local population related to the SSC also declined as a response from 9 percent to 3 percent across this period. For no other category was there a statistically significant shift in the proportion offering a given response.

On the matter of perceived disadvantages, the proportion offering the response of "none" or indicating that they did not know of any increased slightly from 41 percent to 50 percent across this period. For the responses that were volunteered in the original survey, only one involved a significant change in frequency. Approximately one-quarter (27 percent) of the 1987 respondents indicated a concern about environmental harm of one kind or another. In the 1988 survey, this figure was one in ten (11 percent) (see Figure 3-5).

Concerns about radiation and issues related to nuclear power were mentioned by 4 percent of the respondents in the original survey. In

FIGURE 3-4
PUBLIC OPINION SURVEY
PERCEIVED BENEFITS OF THE SSC
LOCAL RESPONSES IN 1987 AND 1988

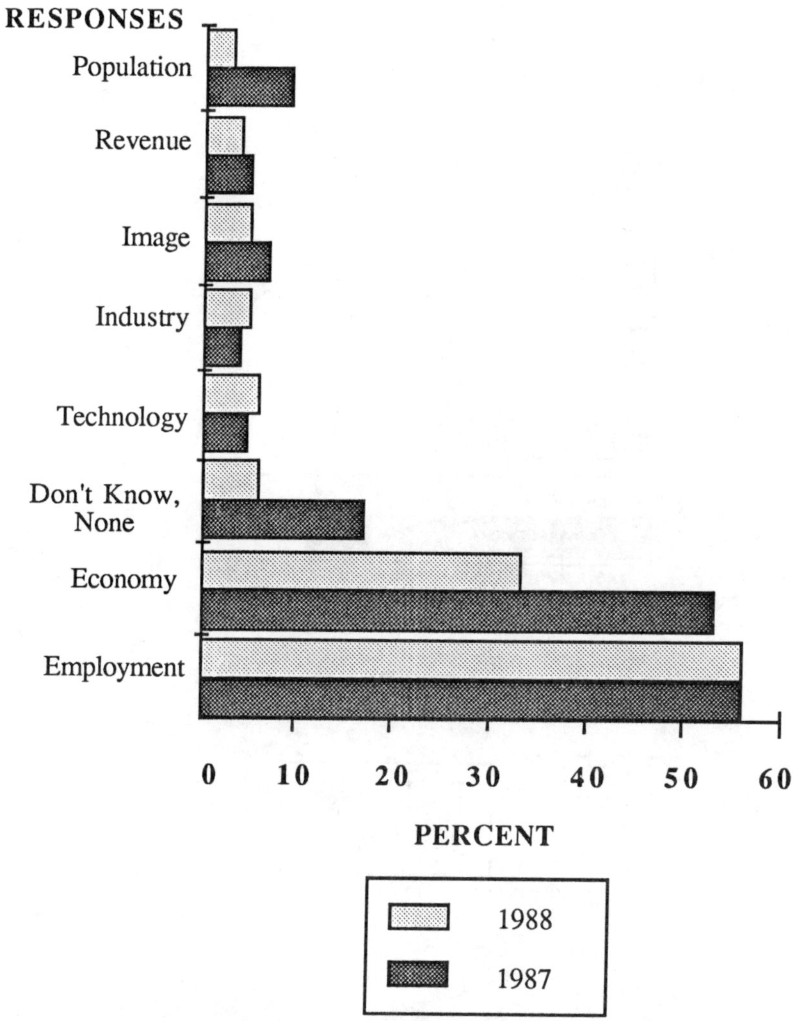

FIGURE 3-5
PUBLIC OPINION SURVEY
PERCEIVED DISADVANTAGES OF THE SSC
LOCAL RESPONSES IN 1987 AND 1988

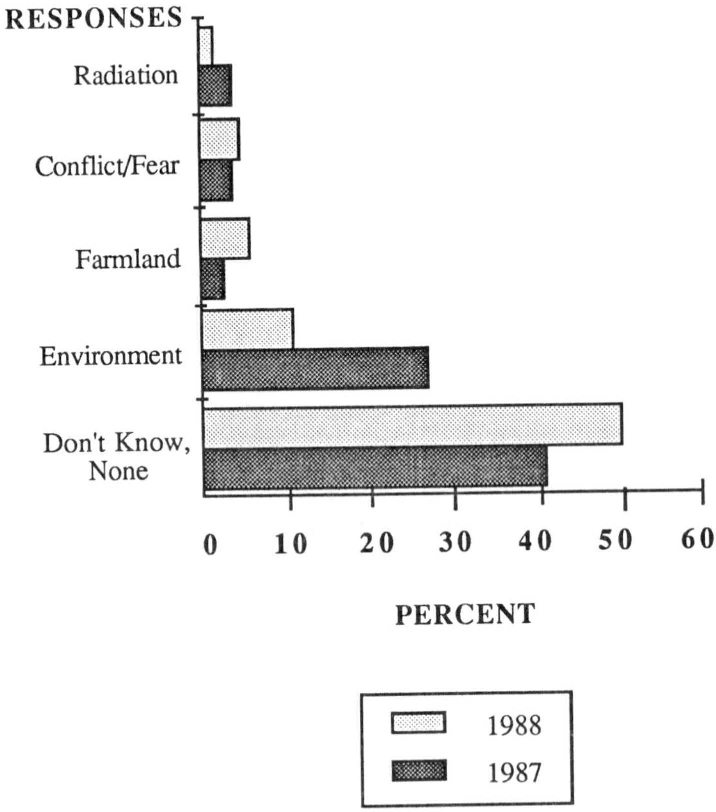

the 1988 interviews, only 2 percent mentioned these issues. This is not a statistically significant difference and should be interpreted as no change in the originally low levels of concern about these issues. This stable trend might be attributed to the availability of the local university-based physics community for answering questions about the SSC and radiation-related issues.

There were two responses which appeared in the 1988 survey for the first time, even though at a very low level. These included concerns about traffic in the area (2 percent), and problems of construction (1 percent).

CHAPTER FOUR

Ethnography Of Land Acquisition

4.1 Introduction

This chapter presents an ethnographic assessment of local responses to the potential impacts of the SSC land purchase and relocation programs in the Stockbridge area. The ethnographic analysis differs significantly from the survey analysis presented in the previous chapter. In the public opinion surveys, telephone interviews were conducted with two random samples, one composed of residents of the State of Michigan and the other composed of Ingham and Jackson county residents (excluding Lansing). In contrast, for the in-depth ethnographic interviews, township officials or clerks were consulted to select respondents. In 1988 a representative group of 57 people who live on the path of the SSC ring and will be impacted by land purchase and relocation were interviewed, with each interview lasting about two hours.

This chapter is organized into six sections, each of which discusses research conducted in 1988. The first and second discuss research goals and methods respectively. The third compares ethnographic findings from 1987 to 1988. The fourth section discusses the "cultural landscape," defined by values people have attached to their land, homes and farms. The fifth section discusses issues regarding land acquisition and relocation. The final section presents a summary of attitudes toward the SSC of people who live on the path of the SSC ring.

4.2 Research Goals

The first research goal was to assess the stability of the 1987 ethnographic interviews in the Stockbridge area by comparing them with findings from 1988. Findings from the interviews that were conducted in 1987 were included in the State of Michigan's proposal submitted to the National Academy of Sciences and the National Academy of Engineering (State of Michigan 1987a).

The second research goal was to understand the concerns raised by people who will be expected to sell land to the State or relocate if the SSC is sited in the Stockbridge area. The needs of these people and how they expect the State to respond to their concerns is addressed. Structured ethnographic interviews were conducted in 1988 to elicit these responses. The data have been presented to the State at timely intervals to increase the likelihood that the land acquisition and relocation programs are acceptable to Stockbridge area residents.

The third research goal was to understand more fully the special impacts that SSC land acquisition will have on approximately 700 property owners, and to determine how property owners wish to be approached by the State of Michigan about the purchase of their property. This analysis was used in April 1988 to guide Gilbert/Commonwealth Inc., the contractor that will approach Stockbridge area residents regarding land purchase (see Appendix C).

The fourth research goal, was to illuminate the potential impacts of the SSC on people who do not live or own property on the SSC ring. This goal will be met in May 1988. Based on social impact assessment research conducted in 1987, small-business owners and local government officials in the Stockbridge area participated in this broader analysis (State of Michigan 1987b). Focus group interviews were to be conducted to identify impacts and to allow participants to make specific recommendations for mitigation to the State of Michigan. Participants also discussed how to increase potential benefits of the SSC among local residents (Caulder 1977; Morgan and Spanish 1983).

This chapter addresses the first two goals. The third goal is addressed in Appendix C which includes the key issues shared with Gilbert/Commonwealth Inc. that will be relevant when property owners are contacted. The fourth research goal will be addressed at the end of May after focus group interviews are conducted to gain an understanding of how the SSC will affect the Stockbridge community.

4.3 Research Methods

4.3.1 Ethnographic Sampling Units

The land acquisition ethnography is based upon a series of in-depth interviews that were conducted in two phases of research; January 15 to February 12, and February 13 to March 11, 1988. Using a methodology that drew upon the knowledge of township officials, 57 respondents were identified and contacted. These 57 people are representative of the 700 that have homes on the path of the SSC ring and will be required to sell surface land, subterranean land or both to the State of Michigan. People who will sell surface land, fee-simple-title, to the State will relocate their homes, farms and businesses. People who will sell subterranean stratified fee estates to the State will not be expected to relocate, unless access to a water supply is blocked. Otherwise, they may continue farming and similar land uses.

A subset of 12 of the respondents were people who had been interviewed during the 1987 ethnographic study in the Stockbridge area. These 12 people were reinterviewed in 1988 to determine whether their attitudes toward the SSC project have shifted.

4.3.2 Consultation With Key Cultural Experts

The method for respondent selection relied upon local township officials who are knowledgeable about the project and the types of people who may be impacted by land acquisition and relocation programs. Because township officials are elected, it was felt that they would be attuned to the people in their communities and the values that govern their lives.

There are 12 townships in the SSC ring area: (1) Stockbridge, (2) Waterloo, (3) Henrietta, (4) Leoni, (5) Blackman, (6) Rives, (7) Tompkins, (8) Onondaga, (9) Leslie, (10) Vevay, (11) Ingham, and (12) White Oak (See Figure 4-1). One person from each township was interviewed, including eight township supervisors, threetownship clerks and one township board member. These key cultural experts suggested a number of types of people who would be affected by SSC land purchase and relocation.

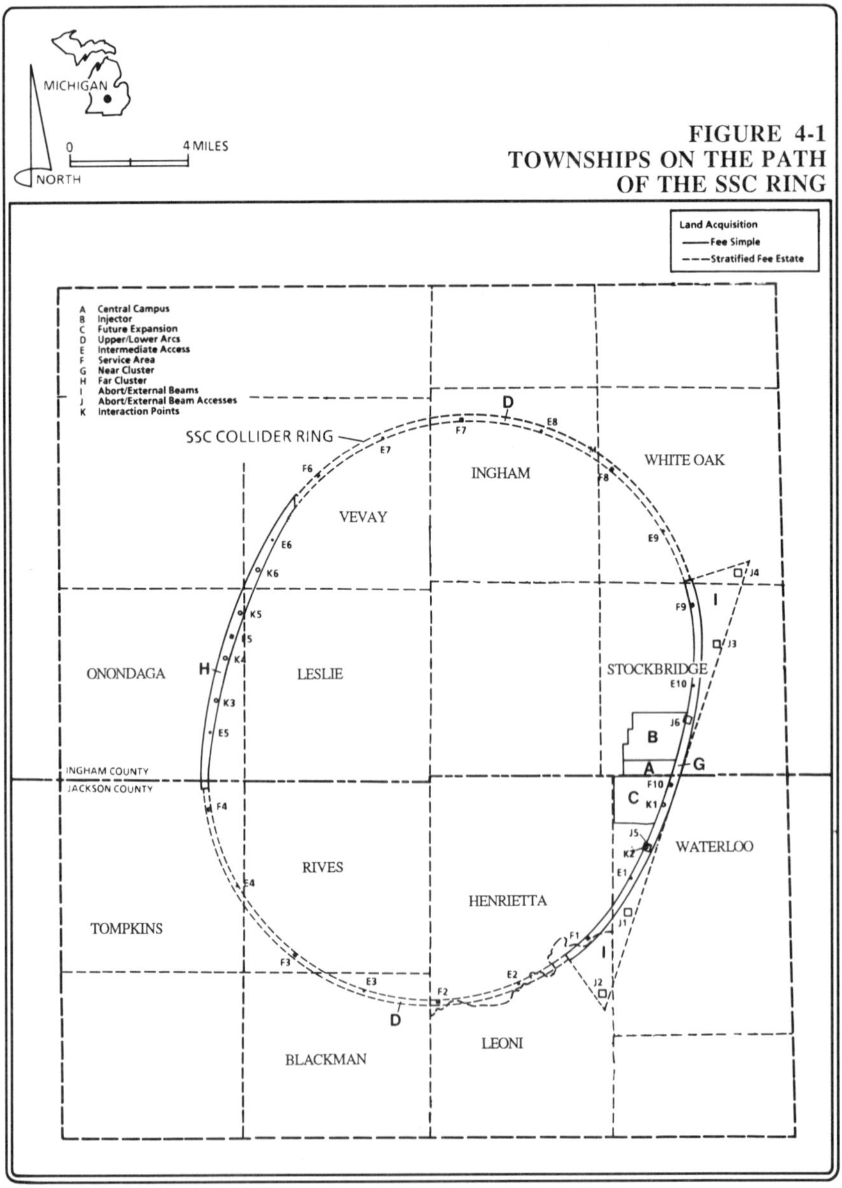

FIGURE 4-1
TOWNSHIPS ON THE PATH OF THE SSC RING

Gilbert/Commonwealth Inc. of Michigan

4.4 Comparison Of 1987 And 1988 Ethnographic Findings

The DOE recommended that ISR researchers determine if attitudes toward the SSC had changed between 1987 and 1988 (Salgado 1988). The 1987 ethnographic study in the Stockbridge area enabled the researchers to perform this comparative analysis.

In 1987, 55 in-depth ethnographic interviews were conducted with people who may be affected by the SSC project if it is sited in the Stockbridge area. In order to ascertain whether there has been a shift in response to the project among these people, reinterviews were conducted with 12 of them, or 21 percent of those who had been interviewed in 1987. These people were classified into four types: (1) large-scale corporate farmers, (2) small-scale family owned farms, (3) rural residents and (4) small-business owners. The people who were reinterviewed own land on or adjacent to the SSC ring that potentially will be sold to the State.

The method used to measure change in attitude toward the SSC involved building a scale to indicate overall response to the project. Data presented in Table 4-1 show assigned scores for attitudes toward the SSC. The overall attitude scores in 1987 were assigned by the ethnographer based on a review of answers.

Table 4-1 demonstrates that attitudes became more positive in every group from 1987 to 1988. The mean attitude score for three large-scale farmers increased from 3.7 to 4.3; the attitude score of four small-scale farmers increased from 2.3 to 3.5. Similar changes in the attitude scores of rural residents and small-business owners show that their attitudes also became more positive over time.

4.5 The Cultural Landscape

In 1988, the ethnographic interviews included questions about the cultural landscape in the Stockbridge area. This information was obtained in order to understand more fully how the State's land acquisition and relocation programs would impact people who live on the path of the SSC ring. "Cultural landscape" describes human modification or alteration of natural surroundings (Rubenstein and Bacon 1983). Cultural landscapes are constructed by people who interact with their surrounding natural environment over generations of time (Wagner 1962). Physical and social attributes are important parts of the cultural landscape. For the purposes of this report,

TABLE 4-1
COMPARISON OF ATTITUDES TOWARD THE SSC IN 1987 AND 1988

TYPES OF PEOPLE	NUMBER (n=12)	ATTITUDE SCORE 1987	ATTITUDE SCORE 1988
Large-Scale Farmers	3	3.7	4.3
Small-Scale Farmers	4	2.3	3.5
Rural Residents	3	3.0	3.7
Small-Business Owners	2	2.0	3.5
Average Attitude		2.8	3.8

Attitude Scores: 1 = Very Negative, 2 = Negative, 3 = Neutral, 4 = Positive, 5 = Very Positive

The exact question wording is:

Given the sacrifices and gains you may experience during the project, how do you feel about the proposed SSC project?

physical attributes of the cultural landscape include aesthetic and environmental qualities, while social attributes of the cultural landscape include heritage values and collective farming strategies.

4.5.1 Values Linked With Cultural Landscape

The following list of values describes how people in the Stockbridge area define their cultural landscape. Aesthetic and environmental values indicate how they relate to the physical characteristics of the cultural landscape. The heritage and collective farming values indicate how they feel about the social characteristics of the cultural landscape.

This list includes responses to questions asked in the 1988 interviews about social and physical values related to the cultural landscape. The values listed here are followed by the number of respondents who mentioned them in the course of the interviews. "No comment" means that respondents did not reply when asked the question. "None" means that there were no values of the kind referred to in the question. The total number of responses exceed 57, the total number of respondents, because a respondent may have mentioned more than one value in the same category.

Case studies illustrate each value, and demonstrate how Stockbridge area residents feel about the physical and social values they have linked with their cultural landscape. The case studies are paraphrased from fieldnotes written during interviews, or directly quoted from transcriptions of tape recorded interviews.

Physical Characteristics Of The Cultural Landscape

This section lists the aesthetic and environmental values associated with the cultural landscape. Aesthetic values are associated with the land's natural beauty found in woods, streams, hills or open spaces. Environmental values are associated with soil, water and air quality. Wildlife for consumption is also an environmental value.

Aesthetic Values

Respondents reported the following values: (34)

Forest wildlife such as deer, rabbits, foxes, quail and pheasants. (16)

Open land and country living. (15)

Ponds with birds, sandhill cranes, muskrats and fish. (9)

Paths through forests used for recreation (skiing, running, hiking). (7)

Woods that provide peace, quiet and privacy. (6)

Landscaping that includes trees, shrubs, and flowers. (3)

"Soft rolling hills." (1)

Stream was main reason bought the farm. (1)

Farm family wants the area to remain the same, have been against other development projects in the area. (1)

Have planted thousands of trees on farmland to make more attractive for home developers. (1)

Respondents reported no aesthetic values: (10)

Respondents had no comment: (13)

Aesthetic Values Case Studies

One livestock farmer said he loves the way the land looks with the "soft rolling hills" and loves the ponds, swamps and woods on his land. He enjoys the deer, pheasants, rabbits and other wildlife that contribute to the aesthetic quality of his land. The beauty of the land is a result of an ecosystem in which all elements of nature are interrelated and should be treated with care.

Another rural resident described her feelings about the aesthetic value of the land:

> We bought this place about nine years ago. We answered an advertisement in the Sunday newspaper. It was actually more

land than we wanted...but we came out and we liked it so well that we bought it. The road was muddy at the time, this house had been rented for 12 years. Very, very poorly kept up. So it wasn't attractive in the usual sense, but I really loved the two hills behind the swamp and the woods. We have six acres of pure woods. We saw it several times that spring and...about the time we decided we wanted to buy it, there were spring beauties covering the path to the forest. We walked over the land several times and each time I liked it more. Buying it meant a 36-mile commute to work, so it was really very much the aesthetics of the place and the possibility of having more space on the farm that convinced us to buy the farm.

Aesthetic Values Conclusion

Over half of the respondents (n=34) described aesthetic qualities of their land or surrounding land. All types of respondents, including mobile home residents, appreciate the wildlife, woods and open spaces that characterize their cultural landscape. Less than half (n=23) of the people interviewed had no comment or felt no attachment when asked how they felt about the aesthetic quality of their rural environment.

Environmental Values

Respondents reported the following environmental values: (20)

Concern for change in well water. (11)

Concern for change in water table. (10)

Pollution and water loss that could disrupt a natural pond or wetland. (6)

Agricultural fields tiled, concerns for proper drainage. (6)

Hunt deer and other animals for food. (6)

Muck lands are important and environmentally sensitive. (5)

Woods are environmentally sensitive. (3)

Farm family is against SSC because they fear it will destroy the earth. (1)

Previous seismographic testing caused water to taste and smell bad. (1)

Unique potential as a bird watching station. (1)

Respondents reported no environmental values: (30)

Respondents had no comment: (7)

Environmental Values Case Study

One family in the Stockbridge area lives almost entirely off their land. They grow produce in a one-acre garden, gather wild strawberries and blueberries, and harvest nuts from their black walnut and hickory trees. They hunt rabbits and pheasants on their land and are permitted to hunt deer on a neighbor's farm. They catch fish from the nearby river. They freeze surplus food from all of these sources.

To this young couple, even more important than living off the land is the quality of life that their present surroundings afford them. Their kitchen overlooks 700 acres, and they often observe deer and sandhill cranes. The solitude and peace and the thrill of observing nature on a daily basis are important to them. They are angered by the idea that trees, plants and their home will be bulldozed. They do not believe the State will or can recompense them for the loss of their home environment.

Environmental Values Conclusion

About one-third of the respondents (n=20) reported concerns for the environment in the Stockbridge area. Most of the environmental values were associated with water: the water table, wells, wetlands, and drainage of land. Some respondents reported that they value the wildlife in the environment as a source of food. Other respondents expressed concern that the woods, mucklands, and the earth in general are environmentally sensitive. One respondent stated that a university has expressed interest in using his woods as a bird watching station. Thirty respondents reported that they did not have environmental concerns for the land; including six who reported that the mucklands are environmentally sensitive, but they would not be concerned if they are drained. Seven respondents had no comment.

Social Characteristics Of The Cultural Landscape

This section discusses the social characteristics of the cultural landscape defined here as heritage and collective farming values. Heritage values are associated with farmland and homes that are passed down through family generations. Collective farming strategies form economic and social values associated with kin and neighbors farming together.

Heritage Values

Respondents reported the following heritage values: (31)

Want to pass property on to children. (23)

Live in centennial home built by hand with pegs, no nails used. (1)

A centennial farm has been in the family for more than 100 years. (1)

Centennial farm is 144 years old. To pay a debt, part of the farm was lost temporarily after the Civil War, but was soon regained. (1)

Has owned 150-year-old heritage farmhouse for 33 years. Intends to pass the home on to children. (1)

Farm has been in the family for 75 years. (1)

Family owned the farm since 1910, sons raised and want to live here. (1)

> House was built in 1853, the farm family is the second owner. They perceive they have a "caretaker " role to research, restore and maintain the home in order to keep its original character. (1)
>
> Old home and they are only third family in it. Feel proud of heritage. (1)
>
> Range of years owned property: 1-20 years (28); 21-40 years (12); 41-60 years (8); 61-80 years (4); 81-100+ years (5).

Respondents reported no heritage values: (25)

Respondents had no comment: (1)

Heritage Values Case Study

One rural family lives in the paternal grandfather's centennial home. The owner said living in his grandfather's home was a "dream come true" and he has many childhood memories of spending time with his grandfather at the house. His grandfather did the farming until he was too old to do so, and now the rest of the family farms the land.

The owners have strong feelings for the land and for the centennial home that was built by hand with the use of pegs, rather than nails. The owners said "the State could not put a price on the house" and that they would "never dream of selling the house." They plan to keep the centennial home in the family and pass it down to their children. The surrounding acreage is farmed by other members of his family. The family enjoys the aesthetic quality of the land and year-round recreation at their pond.

The owners do not want to sell the house and said the State of Michigan should protect the house. If it is on the path of the SSC, the State should offer to move it to "a comfortable distance" from the SSC ring to the other side of the family property. They do not want to be neighbors to the federal government. They are frustrated by the lack of information they have on the SSC project and said the people feel "left out" of the process.

Heritage Values Conclusion

Respondents reported that their farms or homes had heritage value based on one or more of three criteria: the plan to pass the property on to their children, the property has centennial status, or the property has historical value.

Most of the respondents (n=23) reported that their home or land has heritage value because they planned on passing the property down to their children. The SSC land acquisition and relocation programs could threaten the heritage value of these properties.

Persons who own centennial homes or farms (n=3) expressed attachment to the heritage value of their property that they inherited from their ancestors. The case study describes a centennial home owner's attachment to the heritage value of his property.

Some of the respondents (n=5) said that they live in homes that have historical value but do not fit the criteria of centennial homes. They stated that the homes have heritage value and would be passed on to their children and future generations.

The range of the number of years of property ownership in the Stockbridge area indicates that over half of the respondents (n=30) have owned their property from one to 20 years. This group includes the majority of mobile home residents who have on the average owned their mobile homes for eight years. Eleven people have owned their homes and farms from 21 to 40 years; seven people from 41 to 60 years. Eight respondents reported that they have owned their farms and homes from 61 to more than 100 years. Persons in each range reported that their homes and farms have heritage value.

Twenty-five respondents reported that their homes do not have heritage value. These included a few farmers who reported that their children do not farm so the farm will not be passed on to future generations. Other respondents reported absence of heritage value because they do not have children or did not inherit their property from their parents.

Collective Farming Values

Respondents reported the following collective farming strategy: (13)

Shared Help or Equipment With Family: Son (or father) comes in from neighboring farm to help at plowing and harvest times. They share meals and the women help each other care for the children. (6)

Shared Help or Equipment With Non-Family: Share labor and equipment with neighbor. (6)

Shared Land With Family: Farms for or with other farmers who are family. (4)

Farm retired parents' land. (3)

Shared Land With Non-Family: Farms for or with other farmers who are not family. (4)

Reported no collective farming values: (43)

No comment: (1)

Collective Farming Values Case Study

Three brothers use a collective farming strategy. They share ownership of all their equipment and rely on each other to keep the farm going. These property owners live on the area of the SSC ring targeted for fee-simple-title purchase and consequently, their kinship based collective farming strategy is threatened.

The brothers farm several hundred acres together in the Stockbridge area. They described their collective farming strategy in the following way:

> The collective farming is our livelihood. For our operation, you could not do it alone. You can't do it with hired help. You have to have your own family. It's more efficient to farm with your own family. Hired help cannot do the farming. They just cannot do it, that's all. Hired help is very, very expensive and it is very difficult to get good farm labor. You can't get people to do it the way you want it done. If I had to have somebody else help with our farm, I wouldn't do it. If one of us were no longer here, I would probably quit farming.

Collective Farming Values Conclusion

Several farmers (n=13) share labor, equipment and land with family or with non-family in the Stockbridge area. These collective farming strategies fulfill economic and social needs in the farm community. Respondents reported that their collective farming strategies may be threatened by the SSC land acquisition and relocation programs. If the participation of family members and neighbors is no longer available, the nature of the family farm will change.

Forty-three respondents reported that they either do not farm or do not farm collectively. One person had no comment, or did not respond to the question.

4.5.2 Summary Of Cultural Landscape

Several respondents reported aesthetic and environmental values in the physical component of their cultural landscape. The majority (n=34) of people who live on the path of the SSC ring noted aesthetic values. Fewer respondents (n=20) reported attachment to the environmental qualities of their cultural landscape.

Several respondents reported values that link them to the social component of their cultural landscape. Over half of the respondents (n=31) reported that they are attached to the heritage value of their homes and farms. They described heritage value as property that is centennial, historically significant or would be passed on to future generations. More than one-half of the farmers (n=13) who participated in the study (n=24) reported that they value and use collective farming strategies. The SSC land acquisition and relocation programs would threaten the structure of these groups because the participation of nearby neighbors and kin would no longer be available.

4.6 Economic Effects On Types Of People Due To SSC Land Acquisition And Relocation

The results of the 1988 ethnographic interviews are presented here to provide the State and DOE with an idea of the complexity of the issues associated with the land acquisition and relocation programs in the Stockbridge area. This section highlights areas of concern that require further attention and therefore emphasizes potential losses and disruptions in the lives of people who live on the path of the SSC ring. The information was used and will continue to be used to guide policy formation relating to land acquisition and relocation programs.

While fair compensation is the goal of the State's land acquisition program, the SSC project represents a complex set of economic choices for property owners. For example, respondents described numerous instances in which they could experience economic losses due to the SSC land acquisition and relocation programs. The following text first presents concerns for potential economic losses shared by all respondents and then presents concerns that are specific to groups of people that characterize the Stockbridge area. These include rural residents, mobile home park residents, crop and livestock farmers, centennial home or farm owners, and small-business owners.

The State of Michigan has developed and implemented several programs to address or mitigate the impact of the SSC on Stockbridge area residents. These programs respond to the concerns for economic loss and are described in Chapter Five. The State also recognizes that not all problems raised by local residents can be solved, and in some cases, the SSC will require that personal sacrifices be made to accommodate the project.

4.6.1 All Property Owners

This section describes the various economic effects that land acquisition and relocation could have on all property owners. For example, relocatees indicated that the assessment of property value is one of the most important issues related to the potential siting of the SSC project in the Stockbridge area. Most of the people, regardless of their attitude towards the SSC, stated that fair market value, or twice the tax value, would not be adequate compensation for their property.

Options Contract

Several Stockbridge area residents who live on the path of the SSC ring expressed the concern that when they sign the options contracts to sell their property to the State, they will lose control of their property, household finances and farm and business incomes. They stated that they are concerned about how long their property will be "tied up" by the options contract that would restrict them from selling any of the acreage. Furthermore, they would hesitate to expand businesses, develop the land in any way, landscape the property, plant certain kinds of crops, remodel or improve homes, or build additional outbuildings.

Most of the respondents stated that their lives have already come to a standstill while they wait for the DOE to select the site for the SSC. Part of this inertia is due to the lack of information about the land acquisition and relocation programs. Property owners feel like they are "hanging in the lurch" as they wait to be contacted by the State about their land. Several respondents reported that the anxiety associated with "waiting" and "not knowing" about the impact of the SSC on their lives has increased due to the recent announcement by the federal government that the "preferred site" for the SSC will be selected in November 1988 instead of July 1988.

Payment For Land

Most of the people who will be relocated said that fair market value does not compensate them for sacrificing the heritage value of their home and the opportunity to pass the land down to their children and future generations. They also believe the State should pay more than fair market value to compensate them for sacrificing their collective farming strategies and the aesthetic and environmental qualities of their land that are part of their cultural landscape (see section 4.5).

Capital Gains Tax

Several of the respondents expressed concern for the capital gains tax they would have to pay on the income they receive for their property. They feel the payment of capital gains tax would be an unfair financial burden.

Mineral Rights

Most of the respondents who live on the SSC ring are interested in maintaining their mineral rights. Several of these respondents said if this were not possible, the oil companies could estimate the mineral worth of the ground and the State could compensate property owners accordingly.

Taxes

Stockbridge area residents who live on the SSC ring, regardless of their attitude toward the SSC, expressed concern that increased property taxes due to higher property assessments would be a long-term impact of the SSC. They also felt that Michigan taxpayers would pay for the lost tax base due to the State's relinquishing of land to the DOE for the Superconducting Super Collider. If that were the case, the SSC would increase their cost of living.

4.6.2 Rural Residents

Several respondents emphasized that they do not want to go into debt or experience an increase in the cost of living because of a change in residence and relocation activity resulting from the SSC project. Sources of the potential increase in the cost of living are discussed below.

Mortgage Rates

Rural residents said that they do not want to pay higher mortgage rates on their new homes. They recommended the State provide low-interest financing to those families who would have to assume a mortgage to buy or build a new home. Many respondents said they should pay the same rate of interest for the mortgages on their new homes that they pay for their present homes.

Some of the property owners who will relocate are currently purchasing their property under a land contract with their parents or other family members. In this case, they do not pay interest on mortgage loans. These families are concerned that without their present land contracts, they will go into debt as a result of participating in the State's relocation program.

Retirement Programs

Some of the rural residents we interviewed are retired or preparing to retire. They have spent several years planning financially for their retirement and consequently own their home, or owe very little on their home. They feel that the land acquisition and relocation programs threaten their household retirement program because changing residence could increase their overall cost of living, or in some cases, separate them from the farmland that they lease to neighbors, thereby reducing their retirement income.

Property Assessment

Rural residents often obtain firewood from the woods on their land to heat their homes in the winter. If they relocate, they do not want to pay heating costs for other sources of fuel. These property owners want to be compensated for losing the economic value of their woods since their heating costs would increase at another residence.

Some of the rural residents expect to be compensated by the State for other kinds of economic land values. For example, rural residents who have fruit trees on their land anticipate being compensated by the State for their economic value since they use the fruit for private consumption and for sale to the neighborhood.

Property Values

Many respondents expressed concern that property values would decrease because of the SSC project. Some of the respondents will sign an options contract, sell their land and homes, yet continue to live near the SSC. These respondents stated they cannot comprehend the long range impact of living next to the SSC. They are uncertain of what the government property will look like, and how "being a neighbor to the federal government" will change the rural quality of life they now enjoy. These respondents feel that changes in the rural environment could lower the value of land adjacent to the ring.

4.6.3 Mobile Home Park Residents

The land acquisition program will uniquely impact the approximately 140 residents of a mobile home park in Onondaga Township (ISR fieldnotes, March 4, 1988). There are 39 occupied sites in the park and all but nine are owner-occupied. The mobile home park presented a unique situation for the SSC Task Force when they were contacting landowners who will be involved in the land acquisition and relocation programs. Owners of mobile homes are targeted for relocation but rent their mobile home lots and therefore are not listed as landowners on Onondaga Township tax rolls. The SSC Task Force made a special effort to add the names of these potential relocatees to the SSC mailing list.

Most of the residents are low- and single-income families and will require special relocation assistance.

Mobile Home Standards

The mobile home park provides a unique opportunity for affordable housing in Onondaga Township. The park accepts mobile homes that would be rejected by other courts because of unacceptable size and age. To be relocated to another mobile home park, these mobile homes would have to be upgraded or replaced and the costs of doing so would be prohibitive to most of the owners.

Some of the mobile home park residents stated that it would also be difficult to find space in a nearby mobile home park, making relocation very difficult.

Cost Of Living

Several of the mobile home park residents stated that they could not afford to live in another mobile home park because they would be unable to pay the required additional fees for their children and pets.

Costs Of Moving Mobile Home

Many of the mobile home park residents will not be able to afford to move their mobile homes. One resident stated that she recently moved her mobile home to the park in Onondaga Township and paid for 24 separate items ranging from the nuts and bolts of moving the mobile home to transferring telephone service (See Appendix D).

Neighborhood

The residents of the mobile home park share a sense of community and enjoy being neighbors. They do not want to lose their neighbors due to the State's land acquisition program. One resident has lived in the court for 14 years and does not want to be uprooted.

Several other residents do not want to be separated from their families who live nearby.

Schools

Parents of the approximately 48 children who live in the mobile home park stated that the local school system is excellent and they do not want to move out of the school district. Land acquisition and relocation activity could cause these children to change schools if there are no mobile home parks with available space in their school district.

Country Atmosphere

Another special feature of the mobile home park is the country atmosphere, created by the tall shade trees, good well water and quiet setting. Several residents stated that they value, but will be at risk to lose, the country atmosphere when they relocate.

4.6.4 Crop And Livestock Farmers

This section highlights the various potential economic losses perceived by crop and livestock farmers.

Fair Market Value

The farmers we interviewed on the SSC ring stated that they should be paid more than fair market value for their farmland for several reasons that are discussed below. However, most farmers reported that the most important reason they should be paid more than fair market value is because it would not reflect the true value of agricultural land. The recent decline in the value of agricultural land makes it an inopportune time to sell farmland.

Crop Yield

Several farmers who sold or leased easements on their property for use by various companies were "stung" by unforeseen problems that cost farmers substantial amounts of money and aggravation.

These problems were caused by survey and construction activity associated with development projects on the land that resulted in soil compaction, crop damage and water and drainage problems. These impacts reduced crop yield and therefore farm income.

Several farmers expressed the concern that the tunneling for and construction of the SSC ring will alter the drainage on their farmland, causing dehydration of soil in some places and pooling of water in other places. Farmers were also concerned that survey and construction activity will cause soil compaction. Each of these impacts would reduce crop yield for several years, and farmers would lose money. Some farmers said that the price they receive for the stratified fee estate or fee-simple-title property should reflect these possible losses.

Farmers on the SSC ring recommended that the State hire a professional who will ensure that the State compensates farmers for damages and financial loss due to the SSC. The farmers emphasized that many problems do not manifest themselves right away. These damages could become evident during construction, or in the future during SSC operation. The farmers want to protect their farm from damages and financial losses. They recommended that farmers be able to document damages caused by the SSC project, so they can be reimbursed by the State in a timely fashion.

Several farmers are concerned that the SSC will have an adverse effect on other aspects of farm production because they fear the water table might drop and consequently interrupt irrigation activities on the farms. This impact would result in loss of farm income due to reduced crop yield. Farmers want to be compensated for interrupted production and loss of farm income caused by the SSC.

Comparable Land

Many farmers targeted for relocation doubt that they will be provided with, or be able to find, comparable land in a nearby area because all of the farmland in the Stockbridge area is already in use. The farmers further stated that their land is tiled and that they have spent several years building up the soil quality to produce high yields; for those reasons, finding comparable land would be very difficult. These farmers feel that they should be compensated for the full economic worth of their land.

Farm At Two Locations

Some farmers who will sell major portions of their farm acreage to the State would prefer that the entire farm be purchased so they can relocate elsewhere and resume farming with a productive amount of acreage. Many farmers stated it would be impossible to farm in two locations because of the expense and loss of time associated with driving the farm equipment long distances.

Maintain Farm Business

Some of the farmers stated that selling their land to the State of Michigan will cause them to go out of business because of the tremendous cost of rebuilding a farm elsewhere. Some of these respondents said whether or not they stay in farming will depend on how fairly they are compensated by the State for their farmland and farm buildings. They are concerned that they get replacement value at today's cost for their farm buildings and more than fair market value for their land. These farmers find it difficult to contemplate building new outbuildings, obtaining comparable land elsewhere, moving livestock and generally believe that it would be, as one farmer said, "impossible to duplicate the business" at another location. The farmers who will go out of business because of the land acquisition program expect the State to pay them equitably for doing so.

Income

Farmers described other potentially negative impacts of the land acquisition program on the economy of the Stockbridge area farm community. Many of the farmers rent and farm their neighbor's land and will lose it as a source of income if they relocate. They cannot count on having access to leased farmland at their proposed new farm location. This loss of farmland, and subsequent loss of income could make it impossible to continue farming at another location.

The land acquisition program threatens the economic security and kinship structure of farm families that rent large parcels of land that are targeted for fee-simple-title purchase. Some of the farm families can continue to farm their land in the stratified fee estate area, but will lose access to land that will be sold to the State. This situation threatens to fragment kinships that jointly work the land.

One family on the SSC ring farms several hundred acres of land, and their kinship of four families depend on that income. For twenty years, the family has rented approximately one-third of the total acreage they farm, and that land will be purchased by the State for the SSC. The economy of the farm will be negatively impacted if they lose access to the rental land. Loss of agricultural land could cause the kinship to separate. Most of the farmland in the surrounding townships impacted by the SSC is already being farmed, so it would be almost impossible to find a few hundred acres of land to rent nearby to replace the land sold to the State.

Farmers on the SSC ring frequently stated that they farm the nearby land of their parents who are now too old to farm. However, they will be unable to continue to farm their parent's land if they relocate and that will negatively impact their family income.

Kindred And Neighborhood Farm Hands

Farmers also stated they may experience economic loss because they will lose their neighbors who help with baling hay, combining soybeans and other agricultural activities.

Farm Efficiency

In the stratified fee estate area, small parcels will be acquired fee-simple-title for SSC service or access stations. If a station is located in the middle of a field, it could create a nuisance for farmers who will have to plow around it in the field. The farmers said that the stations should be built on nearby roads. This would minimize the problem of circling tractors around them, and also minimize the length of their access roads that would cut through the fields.

Property Assessment

Some of the respondents expressed concern that they may experience economic loss as a consequence of the land acquisition program because the State may not recognize the full economic value of their land. For example, some farm families harvest lumber from their woods every 9 or 10 years which contributes to their farm income. Many farm families cut firewood from their woods and heat their homes with wood stoves. They stated that heating with their own wood saves them substantial amounts of money during the cold months. One farmer said instead of using fuel oil, he burns wood as a source of heat, and that saves him about $200 each month. These families stated that if they lose their woods they will be financially burdened by heating bills that will increase their cost of living.

Moving Livestock

Another economic issue is the timing of the actual land acquisition and relocation activity. Livestock farmers stated that they do not want to move their animals during birthing season.

Dairy farmers expressed concern that because of the temperament of dairy cows, moving them will very likely interrupt milk production for approximately one year. One farmer stated, "once you put them on the truck, they're not the same." Dairy farmers want to be compensated for this probable loss of income.

Dairy farmers expressed concern about moving away from their present milk cooperatives where they have engaged in business for several years. They associate economic loss with moving their dairy farm to a new location and selling milk to a new cooperative.

Water Interruption

Several farmers expressed concern for the lives of their cattle and hogs that depend on several thousand gallons of water on a continuous basis. They stated that if their wells are interrupted and water would be trucked to their farms for the animals, the timing of the delivery would be critical to the lives of their animals. Dairy farmers also stated that variation in the quality of the trucked water could interrupt milk production.

Several other people are concerned their wells will be interrupted, and that it may be impossible to drill new wells nearby.

Retirement Income And Savings

Farms in the Stockbridge area are a current and future source of income, often used as a long-term investment for retirement. Crop and livestock farmers do not have the benefit of retirement pensions as do other professionals and instead use their land as an investment for retirement. Stockbridge area farmers are presently in control of their land and have the option to sell parcels at anytime, especially when land values increase and they would make a profit from the sale. This land is their vehicle for supplementing their retirement income. Many farmers stated that they counted on having this economic security for their retirement. If they have to sell their land to the State for the SSC they should be compensated for the growth potential of their land.

In one case, a landowner fears he may experience an economic loss if is not adequately compensated for the development potential of his land. This farmer plans to sell lots on his property for a subdivision and has additional plans to develop his land in other ways. This farmer said that he would need extensive financial compensation for his property because of this growth potential.

4.6.5 Small Business-Owners

Due to time limitations, only six small-business owners were interviewed. The impact of the State's land acquisition program on small businesses in the Stockbridge area will be examined in greater depth through focus groups that will be conducted in May.

Clientele

Small-business operators stated that when they relocate their businesses they will lose their clientele and consequently experience great financial loss.

Resources

The land acquisition program will cause a loss of resources among some small-business owners and consequently have a negative economic impact. For example, maple syrup processing businesses often supplement family income. To relocate and then replace such a business could take 40 years, or one generation, because of the time needed for a new lot of maple trees to mature.

Other small-business owners reported that they sell resources to the public that are derived from their land, such as fill-dirt. By selling their land to the State, they would lose the resources that are the mainstay of their businesses.

Zoning

Small-business owners stated that their businesses are established in their respective communities and zoning regulations in other locations could prohibit them from moving their businesses.

Growth Potential

Some small-business owners said the State should compensate them for the long-term growth potential of their businesses. They may or may not resume their operations depending on the financial compensation they get from the State and depending on the opportunities to start up elsewhere, make profit, and actually stay in business.

Maintain Small Business

Several small-business owners stated that they would be forced to go out of business if the SSC is sited in the Stockbridge area. They said the State should pay them to go out of business.

4.7 Impacts On The Quality of Life

This section highlights several of the non-economic losses mentioned by respondents that could result from locating the SSC at the Stockbridge site.

4.7.1 Economic Effects On Centennial Home And Farm Owners

Some people with centennial homes on the SSC ring have stated that "there is no way the State can put a price" on the centennial home and farm. These property owners expect the State to leave the houses untouched, or at least move the centennial house to a new location. They said the centennial houses should be preserved.

Centennial farm owners said that fair market value would not be sufficient compensation for their land because they would lose the heritage value and agricultural production value of the land.

4.7.2 Change Of Schools

Almost every potential relocatee who has children currently enrolled in school stated they did not want their children to change school districts. Respondents on the SSC ring stated that they do not want to disrupt their children's studies by moving in the middle of the school year.

4.7.3 Sovereignty And Control Of Farmland

Several respondents stated that they felt that their sovereignty was threatened by the State because of the SSC project. A few people do not want to sell their property to the State for the SSC, but will be forced to because, "they (the State) will get it one way or the other," through eminent domain.

Several property owners have problems with the issue of losing control of the land targeted for the land acquisition program. In the stratified fee estate area, some of these people said they would prefer to lease, rather than sell, the lower strata of ground.

Several respondents emphasized that they are very interested in getting the land back when the SSC is no longer in operation. They are interested in acquiring the first option to purchase back the land they will give up for the SSC, including both surface land and subterranean estates. The idea of getting the land back minimizes the negative impact the project will have on their sovereignty and control of their land.

Other farmers have no problem with the idea of relinquishing some of their surface land or subterranean estate to the State for the SSC, as long as they get a good price. They plan to negotiate with the State about the price of their land.

4.7.4 Nuisance Factor (Traffic And Noise)

Several respondents stated that they felt the price the State offers them for their property should reflect the nuisance factor of the project, including noise and increased traffic. They stated that they should be compensated for losing peace and quiet, the aesthetic qualities they value in the rural Stockbridge area (see 4.5 Cultural Landscape).

4.7.5 Health Risk

Several respondents in the stratified fee estate area expressed concern that radiation generated by the SSC project may pose a health hazard to houses located on top of the ring. They asked questions about what kind of radioactive waste will be generated by the project, and exactly where in the ring radioactive material will be located. In addition, some people are worried about dangers to health posed by the magnetic fields created by the magnets in the ring. These respondents expressed the concern that they be compensated for their exposure to health risks associated with the project.

Several of the farmers and rural residents felt that project authorities should inform people about any health hazards to animals and human beings posed by the SSC.

4.7.6 Privacy

Several farmers said that their privacy and independence is threatened because project authorities and construction workers will be using property or access roads on or near their land.

4.8 Expected Gains From SSC Land Purchase

Respondents were asked questions about what they expected to gain from the SSC. These included gains from sale of land, relocating the farm, relocating the home and gains in income. Several people also volunteered responses about how the SSC might benefit the Stockbridge area, the State of Michigan or mankind in general. These comments are included below.

4.8.1 Sale Of Property

Although 57 people were asked what they would gain from selling land to the State, 15 did not respond because they were either mobile home residents or tenant farmers. The question was relevant to 42 respondents, of which two thirds (n=28) said they had "nothing" to gain by selling land to the State. Two respondents said they either did not know or had no comment.

About one-fifth of these respondents (n=8) indicated that they would gain something by the sale. These gains were described as the opportunity to "do something different" or "retire comfortably." One farmer said that his family might be able to "better themselves" and perhaps raise their standard of living. One respondent said that he hoped to gain more than fair market value for his land and out buildings to be able to resume farming at another location.

4.8.2 Relocation Of Farm

The question of what could be gained by relocating the farm was relevant to 19 respondents, most of whom (12) said they had "nothing" to gain. Two farmers had no comment.

Five farmers responded positively to the question and said they expected certain gains from relocating their farm. Positive responses included the opportunity to expand acreage, increase activity, retire comfortably or live in a more convenient location.

4.8.3 Relocation Of Home

Thirty-four respondents would have to relocate their home because of the SSC project. About two-thirds of these people said they had "nothing" to gain by relocating.

Thirteen respondents said they planned on gaining something by relocating their home. The majority of these people (n=12) were residents of the mobile home park in Onondaga Township. These respondents said they hoped that relocating would enable them to improve their standard of living in several ways, including purchasing a new mobile home or a new lot, making repairs on mobile homes, or purchasing a permanent home. Other expected gains included moving to a more convenient location and improving the quality of life by having a better home. One homeowner said that relocating would be a good opportunity to move into a larger house.

4.8.4 Gain Employment From The Ssc

Respondents were asked if they expected anyone in their family to gain employment with the SSC. The majority of people (n=40) responded negatively and one person had no comment.

About one-fourth of respondents (n=15) said they expected someone in their family to be employed by the SSC. One person thought they would eventually get a raise because the SSC would improve business in the area.

4.8.5 Break Even

Some respondents expect to "break even" or receive fair compensation for their property that will allow them to live in homes of comparable quality in a similar setting. Farmers expect to receive more than fair market value for land and to receive replacement value for their out buildings so they can resume farming at a new location.

4.8.6 Volunteered Responses About Gains From SSC

Seventeen people volunteered responses on the positive gains the SSC would bring to the Stockbridge area and State of Michigan. These comments are listed below.

Employment

Respondents stated that the SSC would provide jobs for people in the Stockbridge area and statewide. These respondents said the SSC would provide clean and stable jobs.

Economy

Several respondents said the SSC would be good for the economy of the State and "put Michigan back on the map." They said the SSC would boost tourism for the State. These respondents said the SSC would improve businesses in the community.

Education

Respondents stated that the SSC would improve the quality of education at the local and state levels.

Humanity

Respondents stated that the SSC would host scientific research that would eventually benefit humanity.

Research

Some of the respondents stated that the SSC would produce good scientific research that will benefit mankind. One respondent said the SSC would be good because it would benefit science itself.

Universities

A few respondents said the SSC would be good for the university system in the State of Michigan.

4.9 Analysis Of 1988 Attitudes Toward The SSC

Attitudes toward the SSC of respondents interviewed in 1988 were analyzed and are presented below.

4.9.1 Attitudes By Types Of People

Respondents have been classified by types of people: livestock farmers, crop farmers, rural residents, mobile home residents and small-business owners. The categories were used to test the hypothesis that attitudes toward the SSC will vary by type of person affected. The hypothesis is based on the idea that certain types of people will have reasons that cause them to oppose or support the SSC more than other types of people. For example, it could be assumed that small-business owners would perceive that they have more to gain from the SSC than farmers because of expected economic growth, and that they would report attitudes more positive that those of farmers. However, Table 4-2 indicates that overall, attitude toward the SSC does not vary by type of people, since the average response to this question was neutral. In conclusion, not one group of people perceives that they will experience extraordinary gains or losses from the SSC.

TABLE 4-2
1988 ATTITUDES TOWARD THE SSC
BY TYPES OF PEOPLE

TYPES OF PEOPLE	NUMBER (n=57)	ATTITUDE SCORE
Livestock Farmers	9	3.4
Crop Farmers	15	3.5
Rural Residents	13	3.2
Mobile Home Residents	14	3.4
Small-Business Owners	6	3.7
Average Attitude		3.4

Attitude Score: 1 = Very Negative, 2 = Negative, 3 = Neutral, 4 = Positive, 5 = Very Positive

The exact question wording is:

Given the sacrifices and gains you may experience during the project, how do you feel about the proposed SSC project?

4.9.2 Attitudes By Type Of Land Purchase

Respondents were also categorized by their participation in the State's land acquisition program: stratified fee estate, fee-simple-title and both. The categories were used to test the hypothesis that attitudes toward the SSC will vary by type of land purchase. The hypothesis is based on the idea that people who will be asked to sell land fee-simple-title (surface land) to the State would be more opposed to the SSC than people who will be asked to sell a stratified fee (subterranean) estate. The former category involves relocation and the latter does not, unless access to a water supply is blocked. These categories of the land acquisition program are defined below.

Stratified Fee Estate

People who own land or homes on the path of the SSC ring and will be asked to sell a lower strata of ground to the State of Michigan. This category also includes people who rent land or homes.

Fee Simple

People who own land or homes on the SSC ring who will be asked to sell their land, fee-simple-title, to the State of Michigan for the SSC. This category also includes people who rent land or homes.

Both (Stratified Fee Estate And Fee Simple Title)

People who own land or homes on the path of the SSC ring and will be asked to sell a subterranean estate and parcels of surface land to the State of Michigan for use by the SSC. This category also includes people who rent land or homes.

Table 4-3 shows that attitudes toward the SSC do not vary by type of land acquisition activity. Respondents in each of the three land acquisition categories reported neutral attitudes toward the SSC project.

TABLE 4-3
1988 ATTITUDES TOWARD THE SSC
BY LAND ACQUISITION ACTIVITY

TYPES OF PEOPLE	NUMBER (n=57)	ATTITUDE SCORE
Stratified Fee Estate	8	3.1
Fee Simple	43	3.4
Both	6	3.2
	Average Attitude	3.3

Attitude Score: 1 = Very Negative, 2 = Negative, 3 = Neutral, 4 = Positive, 5 = Very Positive

The exact question wording is:

Given the sacrifices and gains you may experience during the project, how do you feel about the proposed SSC project?

4.9.3 Summary Of Attitudes Toward The SSC

In some cases, respondents felt they may be in a position to experience hardships or financial losses because of the SSC, yet support the project. This support is expressed because respondents believe that the State of Michigan will take responsibility for problems the SSC causes landowners, and compensate them for any losses. In other cases, people felt "very negatively" toward the project, usually because the potential problems they would experience could not be mitigated by the State. People who feel "very positive" toward the project reported that they feel any sacrifice caused by the project is worthwhile because of the potential economic and scientific benefits the SSC would have to their community and the State. In most cases, people who reported a "neutral" attitude toward the project are waiting to see what kind of compensation they get for their land, and what kind of financial compensation the State proposes to offer them for financial loss or perceived sacrifices associated with the project.

4.9.4 Conclusion

The ethnographic study conducted in 1988 indicates that Stockbridge area residents attach value to the physical and social components of their cultural landscape. Respondents acknowledge that the SSC will cause their aesthetic and environmental surroundings to change. The SSC will also threaten the heritage value of their property and their collective farming strategies.

Respondents were categorized by type of people including crop and livestock farmers, rural residents, mobile home residents and small-business owners. During ethnographic interviews, these people raised concerns for the potential economic impacts they might experience due to the SSC. Some of the concerns were shared by all types of people, and other concerns were raised by specific types of people, such as crop farmers or small-business owners. Respondents also reported that they expected the SSC to affect their quality of life. Finally, respondents reported that they do expect the SSC to benefit them or their families, either through employment or scientific knowledge.

In spite of numerous concerns for the SSC, the average attitude of the respondents toward the SSC was neutral. This overall neutral attitude was attributed to the fact that most people were waiting to obtain more information about the State's land acquisition and relocation programs before passing judgement on the SSC project.

Reference should be made here to methods of demonstrating attitudes toward the SSC beyond the scope of the ISR study. These include two petitions that have circulated in the Stockbridge area. These petitions are addressed in Appendix E.

CHAPTER FIVE

State's Response To Land Acquisition And Relocation Issues

5.1 Introduction

Several concerns about the State's land acquisition and relocation programs were raised during ethnographic interviews. The interviews were conducted while the State was in the process of creating these programs; consequently, many concerns have since been addressed. Respondents raised other general concerns about the local impact of the SSC, and the State has made a commitment to respond to these concerns.

The State has been engaged in a process of systematically identifying, analyzing and resolving many of the problems that could occur in the Stockbridge area because of the SSC. For example, in 1986, it was proposed that a "cut and fill" method be used to construct the collider tunnel. In response to concerns raised by farmers at the Dundee site about the effect of cut and fill on the environment, the construction method was changed to shallow-rock tunnelling.

A second problem raised by farmers was resolved in 1987 by the State. The Farm Preservation Act (PA116) provides tax benefits to farmers who enroll in the program and restrict land use to agriculture. Farmers who withdraw early from the contract must pay a penalty. Farmers at the Michigan SSC sites who are enrolled in PA 116 were concerned that if they sell land to the State they would be penalized. The State worked with the appropriate agencies to waive that penalty.

Early on in the project, the State also recognized that some of the problems posed by the SSC can not be mitigated and will require that residents of the Stockbridge area make personal sacrifices in order to accommodate the project.

In 1988, to respond to local concerns for the SSC, the State has continued to implement this process of problem identification, analysis and resolution. The land acquisition and relocation program demonstrates that the State is working to build programs that will be acceptable to residents of the Stockbridge area.

Solutions to problems related to the SSC are not always readily available. In order to address questions about complex issues, the State has formed various types of research committees. The first committee was formed to determine a fair price for agricultural land. A second committee was formed to determine an acceptable formula for options contracts. A third committee was appointed to investigate SSC health and safety issues. A fourth committee has been formed recently to inform local residents about the tax implications of selling land to the State for the SSC. Finally, the State appointed a fifth committee to examine the impact of relocation on residents of the mobile home park in Onondaga Township.

The State has also demonstrated responsiveness to local residents' concerns for the project by providing them with numerous community education meetings, an SSC "hotline", an SSC information office in Stockbridge, and trips to Fermilab, in Batavia, Illinois. Educational brochures on the land acquisition and relocation programs were disseminated in the communities and others are being prepared. In addition, SSC Study Team members have participated in local radio talk shows and television productions. The State has released a series of reports to the local weekly newspapers and interacted with the press on a regular basis.

The State also responded to concerns raised by the public by creating a question and answer handbook about the SSC. Answers to frequently asked questions were answered based on available information, compiled into a handbook and distributed to township supervisors and SSC team members. The handbook was created to help ensure that consistent, accurate information about the SSC is provided to the community on an ongoing basis. The handbook is discussed in Appendix F.

This chapter presents the State of Michigan's responses to concerns that were raised by Stockbridge area respondents during the ethnographic research. The format of the chapter parallels the structure of Chapter Four, Section 4.6, and addresses in the same order concerns for economic effects of the project raised by different types of people.

5.2 Economic Effects On Types of People Due To SSC Land Acquisition And Relocation

The State of Michigan has developed and implemented several programs to address or mitigate the potential economic impacts of the land acquisition and relocation on local residents. These activities are described below.

5.2.1 All Property Owners

This section corresponds with Section 4.6.1. and provides the State's response to concerns for the potential economic effects of the SSC that were raised by all types of respondents.

Options Contracts

The options contracts provide a method for the State to purchase land required for the SSC. The DOE has made available a schedule for the transfer of title of property targeted for the SSC. The ambitious property acquisition schedule specifies that one-fourth of the property will be transferred to the DOE at various times. Before the DOE announces the final SSC site, the State will work with property owners who live on the first quarter of property to be transferred. The State and property owners must reach agreement on the purchase price of property and on the State's exclusive right to purchase the property.

This agreement, or option, is critical to the success of the land acquisition program. The State is in the process of creating an inducement to the option to enhance the likelihood that it will be satisfactory to property owners. All property owners who sell land to the State for the SSC will receive five percent of its appraised value, with a minimum payment of $500, if the owner signs the option agreement within 30 days of the original offer. The option payment will supplement the purchase price of the property. Option payments will remain in effect throughout the entire land acquisition program in fairness to property owners who will be approached after DOE has selected the Stockbridge site. The options program is not final, but the State is working for its acceptance.

Although it is true that property owners will not be permitted to sell their land and homes during the 12-month period that the options contracts are effective, property owners should proceed with home, farm and business improvements. The appraisal values of properties and businesses will reflect these changes. Therefore, the lives of the property owners do not have to come to a standstill.

The State's effort to inform residents about the SSC is a continuous process. Several efforts have been made to educate Stockbridge area residents about the SSC project in general and the land acquisition and relocations programs in particular. These include several community meetings, the provision of an SSC "hotline," an SSC information office in Stockbridge, and other efforts discussed in the above introduction.

Payment For Land

The State is proposing that property owners who sign options contracts within 30 days of the original offer will be paid five percent of the appraised value of their land; a payment above and beyond the purchase price of the property. This payment, in part, is aimed to compensate property owners for some of the inconvenience they may experience because of the SSC. The options program is not final, but the State is working for its acceptance.

Loss of cultural landscape will be mitigated, in part, by the State's land acquisition program. The SSC Commission will provide an ombudsman to help relocated individuals and families take advantage of a wide variety of existing county, state and federal programs. These programs can help ease adjustments to relocation, and include the federal Soil Conservation District tree planting projects, the Michigan Department of Natural Resources wildlife stocking and habitat development programs, and stress counselling for displaced people. In addition, the Cooperative Extension Service has a number of programs to help relocated farmers plan, manage and improve new farms, fields and livestock.

Capital Gains Tax

The international accounting firm, Peat Marwick Main & Co. has been hired to prepare a brochure on the tax implications of selling property to the State for those people who will be required to do so. The State recognizes that how and when to pay capital gains tax is a complex issue. In response to this concern, they have formed a committee composed of tax experts from Peat Marwick to make available such information.

Mineral Rights

In most cases, because of drilling restrictions to protect the integrity of the collider tunnel, the State will purchase mineral rights from the landowner. The value of the gas and oil resources will be determined in the appraisal process. However, it is expected that the appraisal of mineral rights will be of concern only in a few of the properties.

Taxes

The SSC should stimulate steady and controlled growth that would gradually increase the tax base in surrounding communities and eventually benefit the tax payers. The State will pay local townships for the loss in tax base due to large parcels of land being turned over to the Department of Energy for the SSC. The State's financial compensation for lost tax base will continue until the growth resulting from the SSC contributes enough to the local tax base to compensate for that loss.

5.2.2 Rural Residents

This section corresponds to Section 4.6.2 and includes the State's responses to concerns raised by rural residents that they may go into debt or experience a higher cost of living because of the SSC.

Mortgage Rates

The Uniform Relocation Assistance and Real Property Acquisition Policies Act of 1970 (Uniform Relocation Act) provides interest differential payments to compensate homeowners for any increase in interest rates on mortgages.

Each property owner who will relocate will meet with a relocation agent on an individual basis. The relocation agent will provide them with detailed information and counseling on the relocation options available that would be most suitable for them.

Retirement Programs

The Uniform Relocation Act specifies that the housing must be affordable, decent and safe. The legislation prohibits the State from moving families into housing that they cannot afford.

Property Assessment

The appraiser will recognize land improvements such as fruit trees and wood lots that increase property value. Individuals who will be asked to sell land to the State will have the opportunity to inform property appraisers of what they believe to be the true economic worth of their land. They will meet with the property appraiser during the on-site visit, and have at least one meeting with the negotiator who will work with the farmer to reach an agreement on the appraisal.

Property Values

Property values are not expected to decline because of the SSC. Batavia, Illinois, the host of Fermilab, has experienced a slow and steady increase in property values. The SSC campus and surface structures are congenial to the rural setting of the Stockbridge area, and it is not expected that their presence will cause a decline in property values.

5.2.3 Mobile Home Park Residents

This section corresponds to Section 4.6.3 and includes the State's response to concerns for relocation raised by residents of the mobile home park located in Onondaga Township. The State has appointed a committee to determine the best way to meet the needs of these people and to develop a comprehensive program to address the following items of concern.

Mobile Home Standards

The State is aware that some of the mobile homes may not be accepted by other parks and are working to find a solution to this problem that would be acceptable to relocated individuals.

The State has also made a commitment to find space for mobile homes that would be accepted by other parks, if that situation is agreeable to the individual mobile home owner.

Cost Of Living

The Uniform Relocation Act specifies that relocatees must be moved into housing that they can afford.

Cost Of Moving Mobile Home

The State will pay the cost of moving the mobile home: either the actual, reasonable cost of moving, or payments based on a fixed schedule. The relocatee chooses which payment is most suitable. When necessary, moving costs will also cover related expenses, such as transportation to the new home, payments for temporary housing and the storage costs of personal property. To receive full reimbursement, those applying for relocation benefits must have proper documentation of expenses.

The Uniform Relocation Act specifies that the State must find housing that is suitable and affordable for relocatees, so mobile home owners would not be placed in housing that they cannot afford.

Neighborhood

Loss of neighborhood could be an effect of the relocation activity that cannot be prevented or mitigated.

Schools

Change of school could be an effect of the relocation activity that cannot be prevented or mitigated.

Country Atmosphere

Loss of country atmosphere could be an effect of the relocation activity that cannot be prevented or mitigated. However, various state programs will provide the opportunity for landowners to restore existing vegetation at their new locations.

5.2.4 Crop And Livestock Farmers

This section corresponds to Section 4.6.4 and includes the State's response to concerns for the SSC raised by crop and livestock farmers.

Fair Market Value

The Michigan state constitution requires the State to purchase land at fair market value. The fair market value for property will be determined by professional appraisers working under contract for Gilbert/Commonwealth Inc. The appraisals will reflect an overall property value calculated at its highest and best use of any portion of the property.

A proposal is being prepared that provides for equity payments to agricultural landowners as compensation for possible declines in farmland values that may have occurred in the last ten years. This payment would supplement the fair market purchase price and any relocation assistance payments for which property owners may be eligible. Under the direction of the Michigan State University Cooperative Extension Service, a committee is conducting a survey of land prices in the targeted areas of Jackson and Ingham. If this committee finds that agricultural land values have declined, legislation will be proposed to give farmland owners an equity payment to compensate for any lower, current prices. The Cooperative Extension committee will make its recommendation to the SSC Commission before the summer of 1988. Its findings will be used to formulate appropriate legislation to enact the program.

Crop Yield

The State has hired Gilbert/Commonwealth to conduct the survey work for the SSC. Their professionals will survey the land on foot and carry their equipment by hand. The survey teams will not be driving vehicles into the fields unless they have prior permission from individual property owners. This survey method should substantially reduce the risk of damage to crops and farmland. In the event that surveyors do cause damages, the State will compensate farmers.

Members of the Gilbert/Commonwealth surveying team will also contact property owners in person before the survey activity begins.

Property owners may then inform the survey team how the work can be done to minimize interference with farm activity.

In the stratified fee estate area, construction activity will be limited to easement areas and so soil compaction should not occur in farm fields. In the event that construction of the SSC does cause soil compaction, the State will compensate farmers for that loss.

State and federal construction projects have mechanisms in place that provide farmers with the opportunity to file for reimbursement for damages due to survey and construction activity.

Recent geological studies confirm the State's assertion that SSC tunneling will not affect the water table or interfere with wells. Although there is a possibility that SSC construction could cause some short-term dewatering, it is expected that the water table will remain unaffected by SSC tunneling and irrigation activities will not be interrupted.

Comparable Land

The State of Michigan has put into place a committee to address how relocation might affect farmers in the Stockbridge area. This committee plans to provide technical, financial and planning assistance to farmers who will relocate. In addition, the Cooperative Extension Service has a number of programs to help relocated farmers plan, manage and improve new farms, fields and livestock.

The Michigan Department of Agriculture and Michigan Farm Bureau also have agreed to help the State find suitable and comparable land for these farmers. The Agricultural Stabilization and Soil Conservation Office will also help.

The appraisal of property will take into account property improvements such as tiling of fields and quality of soil. Farmers will be compensated for farmland so that they will be able to purchase farmland of comparable quality at another location.

Farm At Two Locations

If the State is required by the DOE to purchase parcels of farmland that render the farm unproductive, then the State will offer to buy the entire farm. Therefore, the farmer will not be forced to farm at two locations.

In the case where the State purchases a parcel of farmland and the remaining parcel is still productive, then the State will pay damages to the extent that the transaction adversely affects the worth of the remaining parcel.

Maintain Farm Business

The Uniform Relocation Act specifies that farmers must receive relocation cost benefits for replacing their buildings. An equity package has been developed to compensate farmers for the recent decline in the value of agricultural land so farmers will receive more than fair market value for their land.

The State's land acquisition and relocation programs are designed to "make people whole." A few farmers may take Eadvantage of the opportunity to sell their farms to the State and start an early retirement. The State's land acquisition and relocation programs, however, are designed to assist farmers with the process of selling their land, moving their farms, taking advantage of program services and resuming all agricultural activities. The State has operated under the assumption that most farmers who sell land, or sell land and relocate their farms, will stay in farming.

The State will be flexible on the schedule for moving livestock, although the State will be under some time constraints to meet the land acquisition schedule specified by the DOE. By working with the livestock farmer, the State will be informed when birthing season will occur, and livestock will be moved at appropriate times to avoid creating stress on the animals. If damages are incurred because of the move, the State will review damage claims on a case-by-case basis.

Income

Farmers who relocate may have to form new relationships with family and neighbors on whom they can rely to help with farm activity. The relocation services available to farmers can help by introducing them to new people. Farm extension and Farm Bureau groups would also help farmers form new social and economic networks.

The land acquisition program will cause some farm families to lose access to tillable acreage that they rent from neighbors. Some farmers who will relocate may be unable to continue to farm their retired parents' land. Mitigation of these impacts are not readily accessible. These changes in farming may be some of the personal sacrifices that residents of the Stockbridge area would be required to make in order to accommodate the SSC.

Kindred And Neighborhood Farm Hands

Farmers who live in the areas to be acquired fee-simple-title will relocate. In some cases, the relocation may cause farm families to separate. This is an unfortunate impact that cannot be readily mitigated.

Farm Efficiency

Gilbert/Commonwealth has stated that the siting of the access and service shafts is flexible within limits. The shafts will be sited as close to nearby roads as possible, to minimize any effect the shafts or surface structures would have on local residents.

Property Assessment

Individuals who will be asked to sell land to the State will have the opportunity to inform property appraisers and negotiators of what they believe to be the true economic worth of their land. They will meet with the property appraiser during the on-farm visit, and have at least one meeting with the negotiator who will work with the farmer to reach an agreement of the appraisal.

Moving Livestock

The State will be flexible about the schedule for moving livestock, although they will be under some time constraints to meet the land acquisition schedule specified by the DOE. By working with the livestock farmer, the State will be informed when birthing season will occur, and livestock will be moved at appropriate times to avoid creating stress on the animals. If damages are incurred because of the move, the State will review damage claims on a case-by-case basis.

Water Interruption

Recent geological studies confirm the State's assertion that SSC tunneling will not affect the water table or interfere with wells. Although there is a remote possibility that SSC construction could cause some dewatering, it is expected that the water table will remain unaffected by SSC tunneling and irrigation activities will not be interrupted. If wells are temporarily interrupted, the State will provide water to farmers on a timely basis. If damages are caused by loss of water, the State will pay the farmer for those damages.

Retirement Income And Savings

The Michigan State University Cooperative Extension Service will recommend an equity adjustment for agricultural land that will encompass payment beyond fair market value. The equity payment is designed to compensate farmers for the recent decline in the value of farmland (see above, Fair Market Value).

Farmland will be carefully surveyed and a professional appraiser will inspect each parcel to determine its fair market value at current prices. The appraiser will consider all improvements to the property, such as houses, outbuildings, land use and potential value for future use. Owners should meet with appraisers to point out improvements and assets that might not be obvious.

A second appraiser will review the recommendations of the first evaluation. Recent sales of comparable properties in the area will also be considered. The final appraisals will reflect the highest reasonable value of the property at the best use of any portion of the property.

The appraisal process and equity adjustments as well as the entire land acquisition and relocation programs are designed to "make people whole," and protect property owners. Selling land to the State should not interfere with farmers' retirement savings and income.

5.2.5 Small-Business Owners

This section corresponds to Section 4.6.5 and includes the State's response to concerns for the potential effects of the SSC raised by small-business owners.

An SSC regional planning district has been established that encompasses local, county and regional planning groups, and unites them in a cooperative effort. The planning district's purpose is to help "manage the good times." Issues the district will handle include zoning, infrastructure and the impact of the SSC on small businesses. Currently, the district is developing a plan, which will be approved by the governor of Michigan and forwarded to the Department of Energy. This plan will address many of the issues raised by local residents who have small businesses on the path of the SSC ring.

Clientele

The State will provide assistance to small-business owners in finding a suitable replacement location and becoming reestablished. The criterion for the suitable replacement location would be one where the community would have needs similar to that of the original, so the property owner would be able to stay in business. However, a displaced business may receive a fixed payment based on income in lieu of a payment for moving and related expenses if it cannot be relocated without a substantial loss of existing patronage. The amount of the fixed payment is based on the average annual net earnings of the displaced business, but may not be less than $2,500 nor more than $10,000 (Michigan Department of Transportation nd). A relocation agent will meet with small-business owners on an individual basis and provide them with information and counseling on their options.

The Uniform Relocation Act requires the State to provide other types of assistance to relocating businesses. The legislation specifies that displaced businesses receive information on federal, state and local programs that offer assistance through loans and other aids administered by the Small Business Administration.

The Uniform Relocation Act requires the State to assist relocating small-business owners with making applications for loans and in securing managerial and technical advice. The State will also be required to assist small-business owners with completing any required application and claim forms (Michigan Department of Transportation nd).

Resources

The State will provide assistance to small-business owners in finding and reestablishing in a suitable replacement location. The criterion for a suitable replacement location would be one with similar resources so the property owner could continue to stay in business.

Zoning

The State will provide assistance to small-business owners in finding a suitable replacement location and becoming reestablished. A suitable replacement location would include one that has the appropriate zoning to allow the small-business owner to continue to stay in business. Representatives of the State could, in some cases, meet with local zoning boards to assist with the process of establishing the business at the new location. The SSC regional planning district will also help business relocatees with zoning issues.

Growth Potential

Owners should receive financial compensation that will allow them to move their businesses and start up elsewhere. The steady population growth resulting from the SSC should enhance the growth potential of small businesses.

Maintain Small Business

The State's land acquisition and relocation programs are designed to keep affected business owners in business. The steady population growth resulting from the SSC should improve the economy of local businesses.

APPENDIX A

July 1987 Super Collider Telephone Survey Questionnaire

This appendix includes the questions asked about the Superconducting Super Collider in the July 1987 Public Opinion Telephone Survey. All questions were asked in both the local and statewide surveys, unless otherwise indicated. The survey methodology and findings are discussed in Chapter Three.

A1. In general, do you feel that science and technology have changed our lives for the better or for the worse?

1. Better
3. Both; Some of each (VOLUNTEERED)
5. Worse
8. Don't know; no opinion

A2. Do you think that new technologies based on scientific discoveries make our lives change too fast, or has the change been about right?

1. Too fast
3. About right
8. Don't know; no opinion

I am going to read you a series of statements and, for each, I would like you to tell me whether you strongly agree, somewhat agree, neither agree nor disagree, somewhat disagree, or strongly disagree.

A3a. In general, the benefits of scientific research have outweighed the harmful results.

A3b. Scientific discoveries tend to break down people's ideas of right and wrong.

1. Strongly agree
2. Somewhat agree
3. Neither agree nor disagree
4. Somewhat disagree
5. Strongly disagree
8. Don't know; no opinion

A3c. Even if it brings no immediate benefits, scientific research which advances the frontiers of knowledge should be supported by the federal government.

A3d. Scientific discoveries are making our lives easier and more comfortable.

A3e. Unless scientists are allowed to study things that don't appear important or beneficial now, a lot of very beneficial things probably will never be invented.

A3f. Basically all scientific discoveries are good things; it is just how some people use them that causes all the trouble.

1. Strongly agree
2. Somewhat agree
3. Neither agree nor disagree
4. Somewhat disagree
5. Strongly disagree
8. Don't know; no opinion

B1. Several proposals have been made to change the structure of Michigan's economy by trying to attract new businesses to the state. For example, some efforts are being made by the State Government to attract businesses engaged in high technology industries such as computers and robotics. In general, do you think it is a good idea for the state to try to attract such new businesses or is it a bad idea?

1. Good idea
3. Depends (VOLUNTEERED)
5. Bad idea
8. Don't know; no opinion

B2. In order to try to attract new businesses, the State of Michigan has paid for advertising to emphasize certain qualities of the state and its population that make it a good place to live and work. Do you think that such advertising to attract new businesses is a good idea or a bad idea?

1. Good idea
3. Depends (VOLUNTEERED)
5. Bad idea
8. Don't know; no opinion

B3. In order to attract new businesses, state and local governments sometimes offer to reduce their taxes. Do you think that offering reduced taxes is a good idea or a bad idea?

1. Good idea
3. Depends (VOLUNTEERED)
5. Bad idea
8. Don't know; no opinion

B4. Another thing sometimes done to attract new businesses is for the state to purchase land and offer it free so a company can build a plant or another facility like an office building. Do you think that offering free land is a good idea or a bad idea?

1. Good idea
3. Depends (VOLUNTEERED)
5. Bad idea
8. Don't know; no opinion

C1. Have you heard or read about a device called the Superconducting Super Collider, which the United States Department of Energy wants to build for basic scientific research in the field of physics?

1. Yes
5. No
8. Don't know

C1a. Where did you hear or read about it? (Anywhere else?)

1. Discussion among friends
2. Attended a public meeting
3. Read about it in a newspaper
4. Heard about it on the radio
5. Saw a story on television
6. Read about it in magazines
7. Other - PF10 to specify
0. No further mentions

The Superconducting Super Collider--or SSC for short--is a scientific machine in which very small particles are made to collide with each other, and then scientists analyze the results to improve our understanding of matter and the universe. The proposed SSC would be the largest and most sophisticated device of its kind in the world.

C3. Because of its cost, only one SSC can be built in the United States. Several states have expressed an interest in being selected as the site of the SSC because it will mean jobs and might attract new industry. Some of these jobs will come from the construction and management of the SSC, and others will come from businesses which will locate near it after it is built. The competition between states for the SSC project has been compared to the competition for the General Motors' Saturn plant because states are willing to offer land and other special services in order to attract it. Do you think that the State of Michigan should actively try to get the SSC built here or should it not try to get the project?

1. Yes, should try
3. Depends (VOLUNTEERED)
5. No, should not try

C3a. Should the state offer free land to the Department of Energy on which to build the SSC in order to attract the project?

1. Yes, offer free land
3. Depends (VOLUNTEERED)
5. No, don't offer free land
8. Don't know; no opinion

C3b. Should the state offer to use tax money to build roads and other facilities to make it easier to get to the SSC site?

C3c. Should the state offer to reimburse local governments for lost tax revenues if land is purchased and then donated to the SSC project?

1. Yes
3. Depends (VOLUNTEERED)
5. No
8. Don't know; no opinion

C4. Have you heard or read about any proposed sites in Michigan where the SSC might be built? [IF YES] (Where would that be?)

1. Yes, in Monroe or Lenawee counties, near Dundee/Tecumseh
2. Yes, in Jackson or Ingham counties, near Stockbridge/Milan
3. Yes, in southeastern Michigan, no specific location
4. Yes, in some other place
5. No, have not heard
0. No further mentions

C5. The SSC will be built in the form of a large circular ring about 53 miles in length that is primarily underground. Would you be concerned if it were going to be located near your present home?

1. Yes, would be concerned
3. Depends (VOLUNTEERED)
5. No, would not be concerned
8. Don't know; no opinion

-THE FOLLOWING SECTION APPEARED ONLY IN THE LOCAL SURVEY-

D1. Wherever the SSC is built, there are likely to be important and noticeable effects on the local area and the people who live there. If your area were chosen as the site of the SSC, do you think that construction of the SSC would have a positive effect on the local economy, a negative effect, or wouldn't it have any effect?

1. A positive effect
3. Wouldn't have any effect
5. A negative effect
7. Depends (VOLUNTEERED)
8. Don't know; no opinion

D2. How about the environment? (Do you think that the construction of the SSC would have a positive effect on the environment, a negative effect, or wouldn't it have any effect?)

D3. What about services provided by local government such as schools, water and sewer facilities, and police and fire protection? (Do you think that the construction of the SSC would have a positive effect, a negative effect, or no effect?)

D4. How about the effect on the local economy of the people who will work at the SSC facility after it is built? (Do you think they will have a positive effect, a negative effect, or won't they have any effect?)

1. A positive effect
3. Wouldn't have any effect
5. A negative effect
7. depends (VOLUNTEERED)
8. Don't know; no opinion

D5. Overall, would you favor or oppose the construction of the SSC in your area?

1. Favor
3. Depends (VOLUNTEERED)
5. Oppose
8. Don't know; no opinion

D5a. Would you strongly or not so strongly favor/oppose the SSC?

1. Strongly
5. Not so strongly

D5b. (IF FAVOR) What do you think would be the major benefits of having the SSC facility located in your area? (Any others?)

D5c. Would there be any disadvantages to having the SSC facility located in your area? ([IF YES] What are they? Any others?)

D5d. (IF OPPOSE) What are your major concerns about having the SSC facility located in your area? (Any others?)

D5e. Would there be any benefits to having the SSC facility located in your area? ([IF YES] What are they? Any others?)

D6. If you (or anyone in your family living there) could get a job at the SSC facility, would you favor or oppose the SSC, or wouldn't that make any difference to you?

1. Favor
3. Depends (VOLUNTEERED)
5. Oppose
7. Wouldn't make any difference
8. Don't know; no opinion

D7. If the state offered to buy or relocate your home because the land was needed for the SSC, and you were offered fair market value for your property and compensation for your inconvenience, would you favor or oppose the project, or wouldn't it make any difference to you?

D8. How about if the state needed to buy or relocate the home of a family member or relative on the same terms? (Would you favor or oppose the project or wouldn't it make any difference to you?)

D9. And what if the state needed to buy or relocate the home of a neighbor on the same terms? (Would you favor or oppose the project or wouldn't it make any difference to you?)

E1. Now I'd like to ask a few questions about you. About how much time do you spend watching television on an average weekday?

E2. Do you read a daily newspaper?

1. Yes
5. No

E2a. Which one(s) do you read?

E2b. About how much time do you spend reading newspapers on an average weekday?

E3. About how much time do you spend listening to the radio on an average weekday?

E4. About how much time do you spend reading magazines in an average week?

E5. How interested are you usually in television news stories concerning science and technology - would you say you are very interested, somewhat interested, not much interested, or not interested at all?

1. Very interested
2. Somewhat interested
3. Not much interested
4. Not interested at all
8. Don't know; no opinion

E7. And how interested are you usually in newspaper stories concerning science and technology? (Would you say you are very interested, somewhat interested, not much interested, or not interested at all?)

E8. How interested are you usually in magazine articles concerning science and technology?

E9. How often do you usually talk to others about issues concerning science and technology - very often, every now and then, only rarely, or never?

1. Very often
2. Every now and then
3. Only rarely
4. Never
8. Don't know; no opinion

F1. What is the highest grade of school or year of college you completed?

00-12. Enter years of school
13-16. Enter years of college
17. Graduate work
98. Don't know
99. Refused

F2a. Did you get a high school diploma or pass a high school equivalency test?

1. Yes
5. No

F2b. Do you have a college degree?

1. Yes
5. No

F2c. Have you ever taken a college-level course in biology, chemistry, or physics? [IF YES] (How many?)

00. Never took course
01-95. Enter number
98. Don't know; can't remember

F2d. Have you ever taken any college-level courses about or using computers? [IF YES] (How many?)

00. Never took course
01-95. Enter number
98. Don't know; can't remember

F3. Do you own your home, pay rent, or what?

1. Owns or is buying
2. Pay rent
7. Other - specify

F4. What is the name of the community you live in?

F4a. How long have you lived in your community?

0-96. Enter number of years
97. Longer than 96 years

F4b. Would you describe the place where you live as a farm, a town or village, a city, or a suburban residential area outside a city?

1. A farm
2. A town or village
3. A city
4. A suburban residential area outside a city
7. Other - PF10 to specify

F5. We are interested in your present job status. Are you working now, temporarily laid off, unemployed, retired, a student, (homemaker), or what?

1. Working now; on strike; sick leave
2. Temporarily laid off
3. Unemployed; looking for work
4. Retired; disabled
5. Student
6. Homemaker
7. Other (PF10 to specify)
0. No further mentions

F5a. What do they make or do where you work?

F6. To get a picture of people's financial situation, we need to know the general range of incomes of all people we interview. Now, thinking about (your/your family's) total income from all sources, (including your job), did (you/your family) receive $25,000 or more in 1986?

1. Yes
5. No
8. Don't know

Was it...
F6b. ... $35,000 or more
F6c. ... $50,000 or more
F6d. ... $5,000 or more?
F6e. ... $15,000 or more?

1. Yes
5. No

These are all the questions I have. Thank you very much for your time and your help with our research. The Michigan SSC Commission will be happy to send you a report of its activities, if you want one.

APPENDIX B

February 1988 Super Collider Telephone Survey Questionnaire

This appendix includes the questions asked about the Superconducting Super Collider in the February 1988 Public Opinion Telephone Survey. All questions were asked in both the local and statewide recontact surveys, unless otherwise indicated. The survey methodology and findings are discussed in Chapter Three.

101. In general, do you feel that science and technology have changed our lives for the better or for the worse?

1. Better
2. Both/some of each **(VOLUNTEERED)**
3. Worse
0. Don't know/No opinion

102. Do you think that new technologies based on scientific discoveries make our lives change too fast, or has the change been about right?

1. Too fast
2. About right
0. Don't know/No opinion

I am going to read you a series of statements and, for each, I would like you to tell me whether you strongly agree, somewhat agree, neither agree nor disagree, somewhat disagree, or strongly disagree.

103a. In general, the benefits of scientific research have outweighed the harmful results.

1. Strongly agree
2. Somewhat agree
3. Neither agree nor disagree
4. Somewhat disagree
5. Strongly disagree
0. Don't know/No opinion

103b. Scientific discoveries tend to break down people's ideas of right and wrong.

1. Strongly agree
2. Somewhat agree
3. Neither agree nor disagree
4. Somewhat disagree
5. Strongly disagree
0. Don't know/No opinion

103c. Even if it brings no immediate benefits, scientific research which advances the frontiers of knowledge should be supported by the federal government.

1. Strongly agree
2. Somewhat agree
3. Neither agree nor disagree
4. Somewhat disagree
5. Strongly disagree
0. Don't know/No opinion

103d. Scientific discoveries are making our lives easier and more comfortable.

1. Strongly agree
2. Somewhat agree
3. Neither agree nor disagree
4. Somewhat disagree
5. Strongly disagree
0. Don't know/No opinion

103e. Unless scientists are allowed to study things that don't appear important or beneficial now, a lot of very beneficial things probably will never be invented.

1. Strongly agree
2. Somewhat agree
3. Neither agree nor disagree
4. Somewhat disagree
5. Strongly disagree
0. Don't know/No opinion

103f. Basically all scientific discoveries are good things; it is just how some people use them that causes all the trouble.

1. Strongly agree
2. Somewhat agree
3. Neither agree nor disagree
4. Somewhat disagree
5. Strongly disagree
0. Don't know/No opinion

201. Have you heard or read about a device called the Superconducting Super collider, which the United States Department of Energy wants to build for basic scientific research in the field of physics?

1. Yes
2. No
0. Don't know

201a. Where did you hear or read about it? **(PROBE: ANYWHERE ELSE?)**

1. Discussion among friends
2. Attended a public meeting
3. Read about it in a newspaper
4. Heard about it on the radio
5. Saw a story on television
6. Read about it in magazines
7. Other - **(SPECIFY)**________________________
8. No further mentions

The Superconducting Super Collider - or SSC for short - is a scientific machine in which very small particles are made to collide with each other, and then scientists analyze the results to improve our understanding of matter and the universe. The proposed SSC would be the largest and most sophisticated device of its kind in the world.

203. Have you heard or read about any proposed sites in Michigan where the SSC might be built? **(IF YES, WHERE WOULD THAT BE?)**

1. Yes, in Jackson or Ingham counties, near Stockbridge/Milan
2. Yes, in Southeastern Michigan, no specific location
3. Yes, in some other place
4. No, have not heard
5. No further mentions

204. The Department of Energy has evaluated several proposals submitted by states and chosen a few to study more carefully before deciding where to locate the SSC. The Stockbridge area in Michigan is one of those sites. Do you know the names of any other states where the SSC might be located? [Which states are they?] **(DO NOT READ) [RECORD ALL MENTIONS]**

1. Arizona
2. Colorado
3. Illinois
4. North Carolina
5. Tennessee
6. Texas
7. R mentions some other state(s)
0. Don't know any other states

205. The SSC will be built in the form of a large circular ring about 53 miles in length that is primarily underground. Would you be concerned if it were going to be located near your present home?

1. Yes, would be concerned
2. Depends **(VOLUNTEERED)**
3. No, would not be concerned
0. Don't know/No opinion

301. Wherever the SSC is built, there are likely to be important and noticeable effects on the local area and the people who live there. If your area were chosen as the site of the SSC, do you think that construction of the SSC would have a positive effect on the local economy, a negative effect, or wouldn't it have any effect?

1. A positive effect
2. Wouldn't have any effect
3. A negative effect
4. Depends **(VOLUNTEERED)**
0. Don't know/No opinion

302. How about the environment? Do you think that the construction of the SSC would have a positive effect on the environment, a negative effect, or wouldn't it have any effect?

1. A positive effect
2. Wouldn't have any effect
3. A negative effect
4. Depends **(VOLUNTEERED)**
0. Don't know/No opinion

303. What about services provided by local government such as schools, water and server facilities, and police and fire protection? Do you think that the construction of the SSC would have a positive effect, a negative effect, or no effect?

1. A positive effect
2. Wouldn't have any effect
3. A negative effect
4. Depends **(VOLUNTEERED)**
0. Don't know/No opinion

304. How about the effect on the local economy of the people who will work at the SSC facility after it is built? Do you think they will have a positive effect, a negative effect, or won't they have any effect?

1. A positive effect
2. Wouldn't have any effect
3. A negative effect
4. Depends **(VOLUNTEERED)**
0. Don't know/No opinion

305. Overall, would you favor or oppose the construction of the SSC in the Stockbridge area? Would you strongly or not so strongly [favor/oppose] the construction of the SSC?

1. Strongly favor
2. Not so strongly favor
3. Depends
4. Not so strongly oppose
5. Strongly oppose
0. Don't know/No opinion

305a. **(IF FAVOR)** What do you think would be the major benefits of having the SSC Facility located in your area? **(PROBE: ANY OTHERS?)**

0. Don't know

305b. Would there be any disadvantages to having the SSC facility located in the Stockbridge area? **(IF YES, WHAT ARE THEY? PROBE: ANY OTHERS?)**

0. Don't know

306a. **(IF OPPOSE)** What are your major concerns about having the SSC facility located in your area? **(PROBE: ANY OTHERS?)**

0. Don't know

306b. Would there be any benefits to having the SSC facility located in your area? **(IF YES, WHAT ARE THEY? ANY OTHERS?)**

0. Don't know

-QUESTIONS 307 THROUGH 310 APPEARED ONLY IN THE LOCAL SURVEY-

307. If you (or anyone in your family living there) could get a job at the SSC facility, would you favor or oppose the SSC, or wouldn't that make any difference to you?

1. Favor
2. Depends **(VOLUNTEERED)**
3. Oppose
4. Wouldn't make any difference
0. Don't know/No opinion

308. If the state offered to buy or relocate your home because the land was needed for the SSC, and you were offered fair market value for your property and compensation for your inconvenience, would you favor or oppose the project, or wouldn't it make any difference to you?

1. Favor
2. Depends **(VOLUNTEERED)**
3. Oppose
4. Wouldn't make any difference
0. Don't know/No opinion

309. How about if the state needed to buy or relocate the home of a family member or relative on the same terms? Would you favor or oppose the project or wouldn't it make any difference to you?

1. Favor
2. Depends **(VOLUNTEERED)**
3. Oppose
4. Wouldn't make any difference
0. Don't know/No opinion

310. And what if the state needed to buy or relocate the home of a neighbor on the same terms? Would you favor or oppose the project or wouldn't it make any difference to you?

1. Favor
2. Depends **(VOLUNTEERED)**
3. Oppose
4. Wouldn't make any difference
0. Don't know/No opinion

401. Now I'd like to ask a few questions about you. Do you read a daily newspaper?

1. Yes
2. No
0. Don't know

402. Which one(s) do you read?

0. Don't know

403. How interested are you usually in television news stories concerning science and technology - would you say you are very interested, somewhat interested, not much interested, or not interested at all?

1. Very interested
2. Somewhat interested
3. Not much interested
4. Not interested at all
0. Don't know/No opinion

404. And how interested are you usually in newspaper stories concerning science and technology? Would you say you are very interested, somewhat interested, not much interested, or not interested at all?

1. Very interested
2. Somewhat interested
3. Not much interested
4. Not interested at all
0. Don't know/No opinion

405. How interested are you usually in magazine articles concerning science and technology?

1. Very interested
2. Somewhat interested
3. Not much interested
4. Not interested at all
0. Don't know/No opinion

406. How often do you usually talk to others about issues concerning science and technology - very often, every now and then, only rarely, or never?

1. Very often
2. Every now and then
3. Only rarely
4. Never
0. Don't know/No opinion

501. What was the last class or grade you **completed** in school?

1. None - or grades 1-4
2. Grades 5, 6, 7
3. Grade 8
4. High school incomplete (grades 9-12)
5. High school graduate, grade 12
6. Technical, trade, or business school
7. College, University, incomplete
8. College, University, graduate
9. Graduate school
0. Don't know

502a. Did you get a high school diploma or pass a high school equivalency test?

1. Yes
2. No
0. Don't know

502b. Do you have a college degree?

1. Yes
2. No
0. Don't know

502c. Have you ever taken a college-level course in biology, chemistry or physics? **(IF YES, HOW MANY?)**

1. Never took course
2. Number of courses taken _______
0. Don't Know/Can't remember

502d. Have you ever taken college-level courses about using computers? **(IF YES, HOW MANY?)**

1. Never took course
2. Number of courses taken _______
0. Don't Know/Can't remember

503. Do you own your home, pay rent, or what?

1. Owns or is buying
2. Pay rent
3. Other **(SPECIFY)** ____________________

-QUESTIONS 504 THROUGH 505 APPEARED ONLY IN THE LOCAL SURVEY-

504. What is the name of the community you live in?

504a. Would you describe the place where you live as a farm, a town or village, a city, or a suburban residential area outside a city?

1. A farm
2. A town or village
3. A city
4. A suburban residential area outside a city
5. Other **(SPECIFY)** __________________________

505. We are interested in your present job status. Are you working now, temporarily laid off, unemployed, retired, a student, a homemaker, or what?

1. Working now, on strike, sick leave
2. Temporarily laid off
3. Unemployed, looking for work
4. Retired; disabled
5. Student
6. Homemaker
7. Other **(SPECIFY)**
8. No further mentions

506. To get a picture of people's financial situation, we need to know the general range of incomes of all people we interview.

Is your annual household <u>income</u> before taxes:

1. Under $15,000 (under $288 per week)
2. $15,000 to $24,9999 ($289 per week to $480)
3. $25,000 to $29,999 ($481 per week to $576)
4. $30,000 to $39,999 ($577 per week to $769)
5. $40,000 to $49,999
6. $50,000 to $59,999
7. $60,000 or over
8. Refused
0. No answer/Don't know

601. So that my office can check my work in this interview, may I have your name?

602. 1. Male
2. Female

603. **CHECK THE DAY OF WEEK INTERVIEW WAS COMPLETED.**

1. Mon 2. Tues 3. Wed 4. Thur 5. Fri 6. Sat 7. Sun

604. **CIRCLE NUMBER OF CALL ON WHICH THIS INTERVIEW WAS COMPLETED.**

1 2 3 4

PLEASE RECORD AND VERIFY PHONE NUMBER:

AREA CODE	**EXCHANGE**	**NUMBER**

END OF INTERVIEW. THANK RESPONDENT;
Thank you very much for your time.
Have a nice day (evening).

I HEREBY ATTEST THAT THIS IS A TRUE AND HONEST INTERVIEW.

____

INTERVIEWER'S SIGNATURE) (DATE)

APPENDIX C

Contacting Landowners

April 8, 1988

Mr. Les Walcott
SSC Land Acquisition
Gilbert/Commonwealth Inc.
Jackson Office

Dear Mr. Walcott:

Thank you for the opportunity to share the results of our social impact assessment study. Our research team conducted in-depth interviews with 57 Stockbridge area residents. Concerns that these people raised for the land acquisition program are included in the preliminary draft of our report (see enclosure). I hope this information will be useful to you.

I would also like to share findings from our research that may be useful as you and your survey team contact landowners to gain rights of entry.

First, our research shows that most Stockbridge area residents generally have no problem with allowing surveyors to gain access to their property but do expect the survey team to show consideration in the following ways:

(1) Notify property owners about one week in advance of survey activity.

(2) Compensate property owners for damage to soil (holes, drainage and compaction), drainage tiles, crops, trees and other property.

(3) Avoid interfering with planting schedules and work around other agricultural activities.

(4) Take care that the survey vehicles do not get stuck in wet ground in April.

(5) Remove shards of metal, stakes and pieces of trash.

(6) Respect privacy and treat property owners with respect.

Undoubtedly, these recommendations are not new to your team. I am sure that Gilbert/Commonwealth surveyors will be pleased to oblige the property owners in every way possible.

Sincerely,

Richard Stoffle
Associate Research Scientist

APPENDIX D

Moving Costs For Mobile Homes

List of Expenses

1. Truck
2. Labor
3. Electric cord
4. Sewer
5. Convert from fuel types
6. Water lines, heat tapes, insulation
7. Rent tires
8. Rent Axles
9. Tongue bolts
10. Replace and remove tongue
11. Moving sheds and deck extra
12. Skirting, top and bottom rails, stakes
13. Lumber and plastic for some units
14. Rent - security deposit 1 1/2 x rent
15. If home gets stuck - costs of wrecker
16. Costs of lodging while home is being tore down and set up
17. In winter - antifreeze
18. Shingles and or flashing on double-wides or add-ons.
19. Replacement or repairs on sheds to meet park requirements.
20. Storage for R.V.'s, boats, camper tops etc.
21. Lag bolts on double and add ons
22. New blocks
23. Cost $42.00 to transfer and connect phones
24. Road insurance - needed above homeowners insurance when moving.

Source: ISR Fieldnotes February 9, 1988

APPENDIX E

Voluntary Support And Opposition

Through the telephone surveys and ethnographic interviews, ISR researchers provided Michigan residents with an opportunity to demonstrate community support and opposition to the SSC. However, Michigan residents also used other important methods to voluntarily express their support of and opposition to the SSC. Most of these statements have not been recorded but two petitions have been made available to ISR researchers. The following is a summary of these two petitions, one positive and one negative.

At the Stockbridge DOE Scoping Meeting, 758 people voluntarily signed a petition that expressed support for siting the Superconducting Super Collider in Michigan. These people signed under the following header: "We, the undersigned, mindful of the economic, cultural and educational benefits accompanying the installation of the Superconducting Super Collider, and available to all the residents of the Stockbridge area, hereby indicate our support for this project."

During the Stockbridge DOE Scoping Meeting, some people signed a petition opposing the location of the SSC in Michigan. Currently that petition has 200 signatures under the following header: "We, the undersigned residents of ______ County, State of Michigan, do not want the proposed Superconducting Super Collider built on the project site in Jackson and Ingham counties in the State of Michigan."

Both of these petitions are still in circulation to members of the Stockbridge area, so these numbers are preliminary. The final petitions will be checked for authenticity and submitted for consideration to the DOE in the future.

APPENDIX F

SSC Question and Answer Handbook

A Handbook Of Answers To Frequently Asked Questions About The Superconducting Super Collider

Introduction

The ISR team created a handbook of questions and answers related to the SSC. This handbook was compiled for several reasons:

1. Documentation of concerns raised by the public will enable the State to develop policies or programs to address these issues. For example, analysis of ethnographic interviews revealed that many landowners were concerned about potential penalties for selling their land to the State after enrolling it under Public Act 116. This act provides tax incentives for farmers who keep land as farmland and penalties for converting it to another use. This concern was conveyed to the State, which then determined the appropriate department to respond to these concerns. As a result, the Michigan Department of Natural Resources announced that farmers who sell PA 116 land to the SSC would not suffer any penalties.

2. ISR surveys conducted in 1987 revealed that the more people know about the SSC, the more they tend to support the project. The creation of a data base of questions raised by local residents was helpful in identifying what people did not know about the SSC, and providing them with answers that addressed their concerns. The handbook has been made available to the public through distribution to their township supervisors. The handbook also has been used to create information sheets that address specific categories of concerns. For example, handouts were provided at the DOE Scoping meeting on

February 16, 1988, in Stockbridge. These question and answer sheets addressed several categories of concerns, including general information about the project and specific issues such as construction, radiation, real estate, water and roads.

3. Many people on the SSC Study Team deal directly with the public and are responsible for answering a variety of questions about the project. To assist them with this task, they have requested a reference that would be updated on a systematic basis. Providing them with an indexed handbook of questions and answers about the SSC helps to ensure the accuracy and consistency of answers given to the public.

Collection

Questions and answers in the handbook were obtained from a variety of sources:

Questions

Questions about the SSC were obtained in two ways, those systematically gathered and those volunteered by state and local residents. Systematically gathered questions were those brought up during in-depth ethnographic interviews, focus groups, and as a result of the open-ended questions in the telephone surveys about the advantages and disadvantages of the SSC. Volunteered questions were those raised at public meetings, discussed in newspapers articles, and raised in conversations on the SSC Hot Line, which was established to provide the public with a convenient means to seek information about the SSC.

Questions from public meetings were recorded in several ways. Some meetings were tape recorded, and questions were compiled from the transcriptions. Two other methods were used at the DOE Scoping meeting on February 16, 1988. Two rooms were set up outside the main conference area, where real estate, technical and scientific experts answered questions. Local residents had the chance to write questions on a card, with their name, address and phone number, so that they could be contacted. Also, Stockbridge high school students recorded questions that community members asked the specialists, and indicated the number of times a question was asked. These questions were included in the handbook.

Formation of the handbook began in February 1988. It therefore does not include several concerns raised by people at the Dundee site, which did not make the "Best Qualified Sites List." These questions were site-specific to the Dundee area, including questions about the tile drainage system and concerns about converting the area back into swampland. Other issues raised at the Dundee site that are of interest to residents of Stockbridge area are included in the handbook.

Sources of questions were also recorded so that changes in the kinds of questions could be observed over time, as well as whether certain communities had questions pertaining only to them. The majority of questions had several sources. Questions that were extremely common may have no source indicated.

The original wording of questions was retained in the handbook when possible, but edited for the sake of clarity and brevity. For example, many questions in the handbook were constructed so the answers would address several concerns at the same time. For example, question 2.2.1 in the handbook "What would the SSC site look like?" addresses questions about the length of the tunnel, its depth and the appearance of the SSC facilities on the surface. Many questions were asked about the time schedule for the site selection process, so a timetable was formed (question 2.1.6 in the handbook) to consolidate and simplify the response. The foremost objective of the handbook was to address every concern of interest to the public. Nevertheless, some questions were so specific that they were not included in the handbook. These questions, which were most frequently about a specific piece of property, were answered by the Michigan SSC Study Team members on an individual basis, either by a phone call, letter or personal visit.

Answers

Answers were obtained from a variety of sources. Written documents were a primary source for background information, the site selection process and technical considerations of the SSC. They include publications of the Department of Energy, the Universities Research Association, the SSC Central Design Group and the State of Michigan.

Other questions, usually more specific, were given to appropriate specialists to answer. This procedure included personal conversations with the specialists and a mailing of questions to them. To ensure the accuracy of the answers, an effort was made to obtain more than one source for each. In addition, several people proofread the handbook of questions and answers.

Sources for answers also were recorded to help prevent the spreading of false information. By knowing the background of the source, the reliability could be considered. For example, a document from the SSC Central Design Group was considered more reliable than answers recorded at the DOE scoping meeting, or on transcripts of public meetings, when the speaker was not identified.

Because of the complexity of the project, many of the questions remained unanswered at the time of distribution. Yet when new information is received, the handbook will be updated and redistributed to ensure that current, accurate information about the SSC is available to the community.

Organization

The primary consideration that guided the formation of the handbook was to organize it for easy access so that it could best serve those who need to use it. The handbook was organized by categories that represent the major concerns most frequently asked about, such as "Health and Safety" and "Real Estate." An index with categories and subcategories in the handbook helps to simplify access to it. Also, similar questions under each subcategory are grouped together, if not actually combined. Some questions with related concerns were cross-referenced to provide the user with as much information as possible about a particular item.

Problems And Their Solutions

Most problems arising from the handbook were organizational, although some were related to source reliability, timeliness and trying to make sense of technical jargon to provide clear answers. These problems and their solutions are listed below.

Choosing Appropriate Categories

In order to solve the problem of overlapping categories, the purpose of these categories was considered. Since categories are designed to provide easy access to the handbook, those that address the major concerns were used. For example, there were several issues raised about water, including a) water use of the SSC; b) water and radiation; and c) the effect of the SSC on the community water supply. Although all fit under the general category of "Water," these issues raise three separate areas of concern. Therefore, three separate headings were formed. In addition, the questions were cross-referenced so a reader could find additional information in other parts of the handbook.

Overlapping Questions

Many questions raised more than one concern. For example, question 2.2.11 in the handbook asks "What is the engineering involved with building the abort tunnel through a wetlands? (My land contains Bartig Lake, which abuts against the stratified fee abort area, and I'm concerned about dewatering and radiation contamination of the ground water.)" This question raises several issues, including the engineering process through wetlands, radiation contamination and the effects of dewatering. This question was placed under "Construction" because of the engineering element of the question, which was not previously mentioned in the handbook. The other two issues were already addressed by other questions and answers in the appropriate sections.

Miscellaneous Unclassified Questions

Some questions did not fit neatly into any given category. Rather than having one large collection of questions under a "Miscellaneous" heading, similar questions were grouped and placed under several "Miscellaneous" subcategories. For example, under "Health and Safety" and "Quality of Life" there are two separate "Miscellaneous" subcategories. The one under "Quality of Life" includes questions on the possibility of television reception interference and flight restrictions caused by the SSC.

Handbook Expansion Provisions

As the handbook grew, the categories changed. To accommodate this, no initial numbering system was used, to allow for greater flexibility. Once the handbook was ready to mail to the specialists for their input, a numbering system was added to allow for quick reference. By that time, the handbook was pretty well organized and not many changes were anticipated.

Source Reliability

Several factors were considered when recording the sources of answers. In some cases, the true source of an answer was not known. For example, sources from public meetings often were not identified on the tape recording or transcriptions of the recording. Therefore, the reliability of the source was unknown. Also, some facts changed over time, such as the length of the tunnel, from 52 to 53 miles. And much of the official information in documents from the Central Design Group and the DOE, for example, was general and not site-specific to Stockbridge. That fact needed to be kept in mind when answering questions such as those about the depth of the tunnel, for example. To deal with this problem, the most recent source was cited, and several sources were used to develop answers to each question.

Timeliness

Because answering many of the questions required contact with specialists, the timely release of the handbook was affected. Yet the handbook did not have to be completed before it was released, since it is to be updated when new information is available.

Confusing Answers

Relying on specialists for many of the answers also resulted in some of these answers being rather confusing. Each field of knowledge has its own jargon, which often is not familiar to the average person. Therefore, some of the answers were rewritten, with care taken to remain true to their original intent.

ANSWERS TO FREQUENTLY ASKED QUESTIONS

ABOUT THE SUPERCONDUCTING SUPER COLLIDER

APRIL 28, 1988

Prepared for the State of Michigan

and the SSC Commission by

Richard W. Stoffle
Michael W. Traugott
Paula Drury
Camilla L. Harshbarger
Florence V. Jensen
Michael J. Evans

Institute for Social Research
University of Michigan
Ann Arbor, Michigan

INTRODUCTION

This handbook contains a list of questions and answers about the Superconducting Super Collider. Because of the complexity of the project, many questions remain unanswered. Yet when new information is received, the handbook will be updated to ensure that current, accurate information about the SSC is available to the community.

The handbook fulfills several needs. First, documentation of concerns raised by the public will enable the State of Michigan to develop policies or programs to address these issues; second, Institute for Social Research surveys conducted in 1987 revealed that the more people know about the SSC, the more they tend to support the project. The creation of this handbook is helpful in identifying what people do not know about the SSC, and providing them with answers that address their concerns; and third, many people are responsible for answering the public's questions about the SSC. These people need a resource that will help ensure the accuracy and consistency of answers given to the public.

This handbook was distributed to people responsible for answering the public's questions about the Superconducting Super Collider, such as township supervisors. A preliminary draft of the handbook was presented to the United States Department of Energy on March 15, 1988.

SOURCES OF QUESTIONS AND ANSWERS

Questions about the SSC were obtained in two ways, those systematically gathered and those volunteered by state and local residents. Systematically gathered questions were those brought up in in-depth ethnographic interviews and focus groups conducted by ISR researchers, and those brought up during telephone surveys in open-ended questions about the advantages and disadvantages of the SSC. Volunteered questions arose from public meetings, newspaper articles and the SSC Hot Line, which was established to provide the public with a convenient means to seek information about the SSC.

Answers were obtained from a variety of sources. Written documents were a primary source for background information, the site selection process and technical considerations of the SSC. Such documents included those published by the Department of Energy, the Universities Research Association, the SSC Central Design Group and the State of Michigan. Other questions, usually more specific, were given to appropriate specialists to answer. This included personal conversations with the specialists, and a mailing of questions to them. To ensure the accuracy of the answers, an effort was made to obtain more than one source for each. In addition, the handbook was given to several people to proofread.

ORGANIZATION

The questions and answers are categorized to make it as easy as possible to locate areas of interest in the handbook. An index at the beginning of the document lists the categories and subcategories found in the handbook. Some questions with related concerns were cross-referenced to provide the reader with as much information as possible.

1.0 AGRICULTURE

1.1 SOIL

1.1.1 In regards to agriculture, have there been any studies done on the effects of what construction will do to soil structure, soil compaction, soil texture, and lost production at actual sites? [Harshbarger 1987]

1.2 CROP YIELD

1.2.1 If the SSC adversely affects farmland and reduces crop yields in the future, how will farmers be compensated? [Harshbarger 1987]

1.3 WATER

1.3.1 What will happen to cattle and hogs that depend on thousands of gallons of water on a continuous basis, if the SSC affects the water supply? [Stoffle et al., 1988]

1.3.2 If water needs to be trucked in, dairy cows may stop producing milk. What will be done about this situation? [Stoffle et al., 1988]

2.0 BACKGROUND

2.1 GENERAL

2.1.1 What is the Superconducting Super Collider?

The Superconducting Super Collider is a scientific instrument used to study the basic particles from which all matter is formed. Beams of these particles will be accelerated to great speeds and will be allowed to collide at various points around the ring. Scientists then will study the spray of particles resulting from the collisions. A national team of physicists and engineers is designing the SSC, which will be the largest scientific instrument ever developed. The United States Department of Energy is evaluating seven potential sites where it can be located, one of which is in Michigan. The other sites are in Arizona, Colorado, Illinois, North Carolina, Tennessee and Texas. [Gilbert/Commonwealth Inc. 1987]

2.1.2 Why do we need the SSC?

The United States has been an international leader in high energy physics, and the SSC will be the most sophisticated laboratory in the world for basic research in this field. Such research not only creates new industries and new jobs, it stretches current technology to its limits. Experience has taught us that the better we understand nature, the better we are able to control and utilize the forces of nature for the benefit of mankind. [Gilbert/Commonwealth Inc. 1987; Jones 1988]

2.1.3 Is the SSC a weapons facility? [Q Sheet 2/16/88]

High energy physics accelerators are used purely for research without specific applications, and are not involved in weapons research or production. Like all other high energy accelerator laboratories, the

SSC will be an open lab, very much like a university. Unlike a weapons facility, no security clearance is required of scientists, college students or any others working at the accelerator. It is planned that the SSC will be an international research facility, with contributions and participation by scientists from many nations, including China and the Soviet Union. There are no secrets, and the public is invited to visit at any time. [Universities Research Association, nd; Burdock 1988a; Kane 1988a; Jones 1988]

2.1.4 How will a site for the SSC be selected?

Several states are competing for selection as the site of the SSC. A panel of scientists from the National Academies of Sciences and Engineering evaluated 43 initial proposals. The panel then forwarded a list of "best qualified" proposals to the Department of Energy, which will make a recommendation to the president. Seven sites are currently being considered (see 2.1.1). The DOE expects to designate a preferred site in late 1988 and, after an environmental review process, select the final site in early 1989. Construction should begin almost immediately, and is scheduled to take six to eight years. [Gilbert/Commonwealth Inc. 1987]

2.1.5 Why was the Stockbridge site chosen as a finalist?

The Stockbridge site was chosen in accord with Department of Energy specifications because of its favorable geological conditions, rural setting and general proximity to major metropolitan areas and university research facilities. [Gilbert/Commonwealth Inc. 1987]

2.1.6 What is the timetable for the SSC site selection process? [Q Sheet 2/16/88]

1983: Research and development of the SSC formally begins.

March 1986: SSC Central Design Group submits conceptual plan and cost estimate to the DOE.

January 1987: DOE announces Reagan approved construction of SSC.

April 1987: DOE issues Invitation for Site Proposals.

September 1987: DOE receives 43 proposals, screens them and forwards 36 to the National Academy of Sciences/National Academy of Engineering (NAS/NAE).

December 1987: NAS/NAE makes recommendation of most-qualified proposals to DOE.

January 1988: DOE announces best qualified sites.

February 1988: DOE holds environmental impact scoping meeting in Stockbridge, Michigan, and in communities of other six potential sites.

March 15, 1988: Public comment on significant issue related to the SSC ends. Letters sent to SSC Task Force must be postmarked by this date.

August 1988: DOE issues draft Environmental Impact Statement (EIS). A 45-day public comment period begins.

November or December 1988: Secretary of Energy announces a preferred site.

December 1988: DOE issues final EIS.

Early 1989: Final site announced.

July 1989: Construction begins.

Mid-1990s: SSC completed.

[United States Department of Energy, nd; Gilbert/Commonwealth Inc., 1986:11]

2.1.7 When and where is the location of the SSC, within Michigan, fixed? [Stockbridge Tape 8, 1988:2; Response Card 2/16/88]

The Governor requested that the Michigan Energy and Resources Research Association (MERRA) explore the potential for Michigan to develop a site proposal for the SSC in 1983. MERRA, under the leadership of John Mogk, assembled a task force that has been continuously active since then. This task force identified two attractive sites in Michigan following extensive studies and discussions with geologists, environmentalists, government representatives and political leaders. In early 1987 the Stockbridge site area was defined and the specific final site configuration fixed. A second site in Monroe and Lenawee counties was also studied and proposed to the Department of Energy, but was not selected. [Jones 1988]

2.1.8 How long will the SSC be in operation? [Stockbridge Tape 7, 1988:4]

The SSC budget discusses costs of construction and of operation for 25 years; this is a useful figure for planning purposes. However, earlier accelerators built for research in this area of science are still operating. The Bevatron in Berkeley, California (commissioned in 1954), the Brookhaven accelerator on Long Island, New York (1961) and the European accelerator in Geneva, Switzerland (1960) are all still productive research facilities. Some smaller accelerators have been shut down after 20 or more years to be replaced on the same site by newer, different research facilities. [Jones 1988]

2.1.9 Why would the SSC be decommissioned? [Q Sheet 2/16/88]

As mentioned in 2.1.8, some accelerators have been shut down after 20 or more years of research to be replaced on the same site by different research facilities. If and when the SSC is decommissioned, it will occur at the end of the SSC's service life. Decommissioning will involve removal of any radioactive machine parts and components from the tunnel so that the land may be used in its original capacity, without any restrictions on drilling or any other activity. [Burdock 1988b]

2.1.10 What types of research have benefited from collider research?

[Stockbridge Tape 7, 1988:3]

The accelerator technology developed for high energy physics has found many more general uses. The typical modern X-ray machine is a much refined version of early laboratory vacuum tubes; more powerful X-ray machines are small linear accelerators, closely related to those used in particle physics experiments. The utility of X-rays in diagnostic medicine was obvious from the start, but accelerators have also brought unanticipated bonuses, such as the use of particle beams for treating certain kinds of inoperable tumors. Several laboratories have facilities for treating hospital patients, and now there are also plans to bring the accelerator to the patient. Fermilab and Harvard University are both designing room-sized proton synchrotrons for hospital use. Accelerators are widely used in industry too, for example, in ion implantation and the inspection of materials for structural defects. [Universities Research Association 1987]

2.2 CONSTRUCTION

2.2.1 What would the SSC site look like?

The Superconducting Super Collider's major feature is the accelerator tunnel, which, at the Stockbridge site, would be located about 140 feet below the ground. The collider will be in the form of an oval tunnel about 53 miles in length and about 10 feet in diameter. Because the tunnel will be underground, the enormous scale of the facility will not be readily apparent. The campus area will look like a typical research center, with a central office building, an auditorium, and various support and industrial buildings. Every few miles along the ring, small surface buildings containing such things as support facilities and power supplies will be located. No sign of the system will be seen on the surface of land inside the ring, and activity there will exist as it does today. [United States Department of Energy 1987a; Gilbert/Commonwealth Inc. 1987]

2.2.2 Why is a large, 53-mile tunnel necessary? [Harshbarger 1987; White Oak Tape 4, 1988:2]

The length of the tunnel was determined strictly by technical factors that allow for maximum acceleration of the particles. Specifically, the radius of curvature of the beam is determined by the particle energy and the strength of the magnets. The large radius (and hence circumference) is required for the particle energy of 20 Trillion electron Volts in spite of the unprecedented magnetic field strength of 66,000 Gauss. [Kane 1988a; Jones 1988]

2.2.3 How much land will be required by the SSC facilities? [Harshbarger 1987]

Approximately 16,000 acres of land will be required. This includes approximately 7,500 to 8,000 acres, primarily in two areas. One will be used for a campus consisting of laboratories and office space. The other parcel will be over the injector area and where the experiments will take place. Along the ring, underground easements for another 8,000 acres also will be required. Some of the land to be acquired is for possible future expansion of research facilities. [Gilbert/Commonwealth Inc. 1987; State of Michigan 1987:2-5; United States Department of Energy 1987:2,49; Jones 1988]

2.2.4 How wide is the needed strip of land above the ring? Why is it this width? [Q Sheet 2/16/88; Response Card 2/16/88]

The strip of land above the ring in the "cluster" areas, where fee-simple-title (see 8.3.1) to the land is required, will be between 1,300 and 2,000 feet, in order to allow flexibility in the final design and location of the experimental areas. Along the arcs, a stratified fee estate (completely underground) (see 8.8.1) of 1,000 feet wide is required, both for flexibility in final design and placement and for high standards of radiation safety. [Jones 1988]

2.2.5 How will construction of the underground ring take place? [Gilbert/Commonwealth Inc. 1987] Tunnel-boring machines will be used to construct the site of the ring. Each machine can proceed several miles before material must be sent to the surface, depending upon geological conditions. It is expected that surface access will be required every 2.5 to 3 miles at the Michigan sites; therefore, there will be approximately 25 access points around the circumference of the collider ring. [Gilbert/Commonwealth Inc. 1987]

2.2.6 Where will the excavated soil and rock be placed? [Harshbarger 1987]

Several possibilities are being considered. These include using the soil to create a recreational area, and giving it to the road commission to use for roads. Also, when trucks bring in sand and gravel to make the cement for the tunnel, these trucks could cart away the excavated material to gravel pits that no longer would be used. [Heinzman 1988a]

2.2.7 How much noise and disturbance will there be during construction and operation of the SSC? [Stockbridge Tape 7, 1988:1]

Construction noise for the surface buildings, as in the Campus area, will be typical of construction of any similar commercial or industrial buildings. The noise from tunnel boring deep underground will not reach the surface, but trucks carrying tunnel spoils, will be as noisy as trucks involved with quarrying or highway construction. The four experimental areas may be excavated from the surface and this could also produce noise locally. [Jones 1988]

2.2.8 Are there going to be any trailer parks for construction workers? [White Oak Tape 5, 1988:1]

2.2.9 Is the siting of the access facilities flexible? Would any be built in the middle of a field so that the farmer would have to plow around them? [Stoffle et al., 1988]

There is some flexibility in the siting of access facilities. Every effort will be made to locate them on existing roads for reasons of economy as well as accommodation of farming and other current land uses. [Jones 1988]

2.2.10 Will there be a fence around the tunnel? [Q Sheet 2/16/88]

No, there will be no need for a fence around the tunnel, which will be about 140 feet underground.

2.2.11 What is the engineering involved with building the abort tunnel through a wetlands? (My land contains Bartig Lake, which abuts against the stratified fee abort area, and I'm concerned about dewatering and radiation contamination of the ground water.) [Response Card 2/16/88]

The abort tunnels will be tunneled underground in bedrock and should not have any effect on surface wetlands. The aborts will be engineered to avoid any possibility of contamination of ground water or soils. Experience at existing accelerator facilities demonstrates that this can be done successfully. [Jones 1988]

2.3 TECHNICAL

2.3.1 Why does the accelerator use superconducting magnets rather than conventional magnets?

Superconducting magnets have stronger magnetic fields. If conventional magnets were used to build an accelerator of the same power, the circumference of the ring would be about 180 miles, rather than the 53 miles needed for the Superconducting Super Collider. In addition, a system using superconducting magnets is less expensive to operate since it requires less electricity to function than does a conventional system. [Peter Limon, SSC Central Design, U of M Physics Colloquium, Feb.10, 1988]

2.3.2 At what temperature are the superconducting magnets kept?

The magnets must be maintained at an operating temperature of 269 Celsius degrees below zero. [Gilbert/Commonwealth Inc., 1986:7]

2.3.3 How does the cooling system work?

Liquid helium is used in the refrigeration system because it is the only substance that maintains its fluid properties at the necessary operating temperature. The liquid helium goes from the refrigerators into the magnets at ten locations around the ring, cooling the superconducting coils as it flows through. The liquid helium then is withdrawn and sent back to the refrigerators for recooling. About 19 million liters of liquid helium is used for this process. [Gilbert/Commonwealth Inc., 1986:8]

2.3.4 Are the chemicals used in the cooling process (liquid nitrogen and helium) volatile chemicals? [Stockbridge Tape 7, 1988:1]

Taking "volatile" to mean "evaporating rapidly or easily at room temperature," these chemicals are volatile; yet they are kept in a closed system, unexposed to air, so the question does not apply. If "volatile" was intended to mean "dangerous," these chemicals, when used properly, are not dangerous. [Kane 1988b]

2.3.5 What is the abort area? What is it used for? [Hot Line 1988a]

The site required contains two roughly triangular aprons on one side of the ring for which a stratified fee estate (underground) is required. There are three reasons for these areas: first, the beam absorbers (aborts or dumps) will be located in part of these areas, although they will require only a small fraction of the areas, probably very close to the accelerator tunnel. Second, "test" beams may be developed in these areas; that is, beams can be brought from the machine for development and testing of research apparatus. Third, in the future, some beams may be brought out of the machine for experiments, and underground experimental halls would be built to serve them. The second and third uses are not part of the requested plan and budget, but the land acquisition was designed to accommodate them as future possibilities. [Jones 1988]

2.3.6 What is the external beam access area? [Q Sheet 2/16/88]
The external beam access areas are for the development of surface facilities and vertical access to underground experimental areas for external beams in the event that these would be developed in the future. (see 2.3.5) [Jones 1988]

3.0 ECONOMY

3.1 EMPLOYMENT

3.1.1 How many jobs will be created by the SSC? [Harshbarger 1987]
After creating jobs at the outset for 4,500 construction workers, the SSC will attract a work force of about 3,000 scientists, engineers, technicians and support personnel, including cooks, accountants and librarians. About 2,500 of these people would make up the SSC's permanent work force, and 500 would be visiting scientists on short-term stays ranging from a few days to several months [Gilbert/Commonwealth Inc. 1987; Gilbert/Commonwealth Inc. 1986:14]

For comparison, about 8,200 people work at the European Center for Nuclear Research (CERN) in Geneva, Switzerland, which is the site of several accelerators. About 3,500 are permanent employees, including scientists, engineers, technicians and support staff, while the majority of the remainder are visiting scientists and students. [CERN public information booklet 1987]

Of the 2,500 people that make up the permanent work force, a percentage will be local people who qualify for the types of jobs available at the research facility. This percentage is not known at this time. Fermilab, which is usually considered the closest example of what the SSC will be like, has a program designed to encourage and prepare local people to become employees of the facility. The program involves visits by high school students, internships for college students, and specialized on-the-job training. Although these few observations suggest it is reasonable to expect a similar arrangement for the SSC, formal research is required before predictions can be made about the numbers of people that would be hired and the types of jobs that would be available. [Stoffle et al., 1987:56, 59]

3.1.2 Will it be guaranteed that a fair percentage of local people will gain employment with the SSC? [Harshbarger 1987] (see 3.1.1)

3.1.3 Will a local hire policy be developed and implemented? [Harshbarger 1987] (see 3.1.1)

3.1.4 Will local people receive job training so they are qualified to take SSC jobs? [Harshbarger 1987] (see 3.1.1)

3.1.5 Will there be a job placement list available? [Q Sheet 2/16/88]

3.1.6 What type of unscientific support jobs will be needed? [Response Card 2/16/88]

3.1.7 Do they think that the staff would be from Michigan? [Q Sheet 2/16/88]

Michigan would have just a few representatives from this state. There would be just as many from Michigan as from other places. [Q Sheet 2/16/88]

3.1.8 Who will determine types of jobs, pay scales etc? [Q Sheet 2/16/88]

The SSC will be a Government owned-contractor operated (GOCO) facility like other Department of Energy laboratories and facilities. Other DOE high energy physics labs are operated by a university or a corporate consortium of universities under contract with the DOE. Lab employees are employees of the contractor and are not federal civil service employees. It is expected that the management of the SSC will be similar. [Jones 1988]

3.1.9 What kind of wages will jobs generated by the SSC pay? [White Oak Tape 5, 1988:3]

The SSC will generate a wide variety of jobs, many of which will require training. There will be relatively few low-level wage jobs. [White Oak Tape 5, 1988:3]

3.2 SMALL BUSINESS

An SSC regional planning district has been established that encompasses local, county and regional planning groups, uniting them in a cooperative effort. The planning district's purpose is to "help manage the good times." Issues the district will handle include zoning, infrastructure and the impact on people affected by the SSC. Currently, the district is developing a plan, which will be okayed by the governor of Michigan and sent to the Department of Energy. This plan should address many of the issues raised below. When the information is available, it will be included here. [Burdock 1988c]

3.2.1 Will low-interest loans be made available to local small businesses so they can expand and compete with incoming businesses? [Harshbarger 1987]

3.2.2 Will local businesses be wiped out by national chains that move into the area? [Harshbarger 1987]

3.2.3 Can local business adequately serve the population growth resulting from the SSC? [Harshbarger 1987]

Recall that the SSC staff and visitors will number about 3,000 when the laboratory becomes operational in the late 1990s. This will involve about 10,000, including families. If one-half to two-thirds are people currently living within commuting distance, new residents will number 3,000 to 5,000. They will be spread among Lansing, East Lansing, Mason, Williamstown, Howell, Ann Arbor, Dexter, Chelsea, Jackson, Charlotte, Stockbridge, Leslie et cetera. The combined populations of these towns is perhaps a half a million, so that the impact on the population of the region, about 1 percent, will be modest. This impact will be greater in Stockbridge and Chelsea, probably, but still not overwhelming. [Jones 1988]

3.2.4 Property values will increase and raise the cost of renting out business office buildings. Why should local businesses have to pay for this impact? [Harshbarger 1987]

3.2.5 Can the government or the SSC officials fund a grant writer to work for local governments or local communities? [Harshbarger 1987]

3.2.6 Will small-business owners be advised as to how to maximize potential benefits from the SSC? [Harshbarger 1987]

3.2.7 Will small business owners be advised about the specific kinds of assistance that might be made available to them to mitigate the impact of the SSC on residents of those counties? [Harshbarger 1987]

3.2.8 How can local businesses get the education and learn how to cope with the oncoming economic development and learn how to evaluate the risk and development opportunities? [Harshbarger 1987]

3.2.9 How will local business persons get the education they need to compete effectively with national chains coming into the area? [Harshbarger 1987]

3.2.10 Will local contractors have a shot at managing components of SSC construction? [Harshbarger 1987]

3.2.11 Could a local development entity be appointed to ensure that a local pool of sub contractors available in the area for work do indeed get work with the SSC? [Harshbarger 1987]

3.2.12 Can assistance be provided to local contractors for the administration of federal paper work which will be required for SSC construction projects? [Harshbarger 1987]

The State of Michigan stands ready to provide assistance to local contractors and businesses to share in the construction of the SSC. A procedure is already operative in the Department of Commerce that provides various assistance to small businesses, including application for federal projects. [Burdock 1988c]

3.2.13 Can the SSC bids for construction be broken down into separate components for various building projects such as industrial space, office space and housing? [Harshbarger 1987]

3.2.14 Can a streamlined procedure be developed to expedite payment to local contractors for their jobs done on SSC projects? [Harshbarger 1987]

3.2.15 How will the engineer be selected to design the water and sewer system expansions which will be required at Stockbridge? [Response Card 2/16/88]

3.2.16 What is the role of the union in the construction of the SSC?
The Michigan State Building and Construction Trades Council has made the Department of Energy a no-strike pledge for members who will work on the SSC. [Burdock 1988c]

4.0 ENVIRONMENT

4.1 LAND/WETLANDS

4.1.1 Will the SSC project harm sensitive ecological habitat?
One of the criteria being imposed on the siting of the SSC facilities is the avoidance wherever possible of sensitive ecological habitats, such as wetlands or mature woodlots. The design of the site has been prepared in consultation and cooperation with the Michigan Department of Natural Resources. [Gilbert/Commonwealth Inc. 1987]

An inventory of the land types that may be sensitive or unique was undertaken to specifically address this question. No wilderness or wild and scenic rivers were identified. Sensitive and unique habitats identified were outside the area potentially influenced by construction of the SSC ring. In areas where the location of building sites is flexible, construction on wetlands can be avoided. Wetlands in areas where building placement is not flexible will be appropriately mitigated by the State of Michigan. [Heinzman 1988c]

NOTE: A more detailed discussion of the nature of the sensitive ecological habitats is found in the State of Michigan's March 15, 1988 submittal to the Department of Energy. Information on the subject is included in Appendix D, section D2.1.7 (page 39) and section D2.3.1 (page 89). The original State of Michigan proposal (Sept. 2, 1987) covers these concerns in Volume 5, section 5.8.1.6 and contains a map (in the Map Supplement volume), figure 5.1-1, which identifies the areas discussed. [Heinzman 1988c]

4.1.2 What impact will the project have on the wetland area? [Stockbridge Tape 7, 1988:2]

The DNR is working with geologists and the Department of Energy and the contractors in answering the question of the project's long-term environmental impacts. There is a small wetland area which can be mitigated by having other wetland areas created. If the DNR says that the project will have a great environmental impact, that will be a factor in the proposal to keep Michigan from getting the project. [Stockbridge Tape 7, 1988:2]

Although some loss of existing wetland resources would undoubtedly result from SSC development, those likely to be affected are of local significance only, and are abundant in the area. Specific plans for wetlands mitigation would be addressed as part of the EIS and regulatory/permitting process (Michigan's Wetland Protection Act and state administration of the Clean Water Act, Section 404), and in consideration of recommendations developed under the mitigation policy of the U.S. Fish and Wildlife Service (FWS).

The State of Michigan, through the Michigan DNR, will play a key role in developing wetland mitigation for the SSC project in close cooperation with the DOE and FWS. The Michigan DNR considers the following parameters applicable to wetland mitigation: (a) mitigation will be by replacement of wetlands unavoidably lost; (b) where feasible, functional replacement (i.e., replacement in kind) is considered to be first choice; (c) replacement will normally be on-site or in the immediate vicinity unless State and Federal regulatory review concludes that a more remote location in the general vicinity is preferable, and (d) replacement of wetlands will, to the degree practical and functional, be developed in conjunction with other project activities and plans as long as other above constraints are met. [State of Michigan 1987:5-69]

4.1.3 Will the SSC alter drainage of the land and dehydrate the soil? [Stoffle et. al, 1988]

Effects on any surface soils would be related to the removal of a wetland for construction purposes. These losses will be mitigated and a replacement in kind where appropriate will occur. Temporary disruption of near-surface hydrogeology (which could cause some soils to "dehydrate") may occur in selected areas where construction is necessary. These effects would be temporary in nature, and recovery of soil moisture would occur when construction is completed. [Heinzman 1988c]

4.1.4 What will happen when dewatering takes place in the wetlands area? [Response Card 2/16/88]

If dewatering is necessary to the degree that the wetlands need to be replaced, (i.e. mitigated), it will be addressed under the State of Michigan Wetlands Protection Act. If the effects are temporary in nature, the wetland will return to its previous condition when the water level returns. Wetlands are able to withstand dry conditions for extended periods and recover from those effects. [Heinzman]

4.2 WILDLIFE

4.2.1 Will the cranes be hurt? [Q Sheet 2/16/88]

The surface disturbance resulting from this project will be minimal and completed in a manner that will not influence the cranes. Construction techniques, noise and other human disturbances will be minimized if nesting areas are in the proximity of the SSC construction. Disturbance of sandhill cranes is discussed in the State of Michigan Proposal, Volume 5 on page 5-74 (September 2, 1987). [Heinzman 1988c]

4.2.2 Will construction of the SSC affect the animals? [White Oak Tape 4, 1988:3]

Some of the SSC facilities would displace certain wildlife habitats. Wildlife species affected would include fox squirrel, white-tailed deer, raccoon, fox and cottontail rabbit. Some disturbance of wildlife species from noise and human activity can also be expected during both construction and operational phases of the SSC, including disruption of normal activity patterns or precluding use of adjoining areas by certain wildlife species. Such effects can be minimized with functional noise abatement equipment on vehicles and other heavy equipment, and by restricting activities to planned, designated areas, particularly where natural habitat exists in nearby areas. With these controls, such adverse effects are not expected to be significant. [State of Michigan 1987:5-74]

5.0 GOVERNMENT

5.1 LOCAL, STATE, FEDERAL ROLES

5.1.1 A variety of questions have been asked about the role of different levels of government, local, state and federal. Who pays for the land, who would own the facility, who would set hiring policy and wage scales, who would maintain roads, etc.... A chart that lists the responsibilities of the various levels of government would be very helpful. [Stockbridge Tape 7, 1988:1; Q Sheet 2/16/88]

5.1.2 Who will own and operate the SSC? [Q Sheet 2/16/88]

The SSC will be a Government owned - contractor operated (GOCO) facility like other Department of Energy laboratories and facilities. Other DOE high energy physics labs are operated by a university or a corporate consortium of universities under contract with the DOE. Lab employees are employees of the contractor and are not federal civil service employees. It is expected that the management of the SSC will be similar. [Jones 1988]

5.2 FUNDING

5.2.1 What are the various fundings for the SSC? [Q Sheet 2/16/88]

The SSC will be constructed and operated as a program of the U.S. Department of Energy. Funding will be part of the DOE federal budget, proposed by the administration and approved by Congress. [Jones 1988]

5.2.2 What happens if, in the middle of the project, the government decides it doesn't want the SSC? [Stockbridge Tape 7, 1988:3]

The possibility that the SSC would not receive final funding from the federal government does exist. The funding, however, must be in place before the construction gets underway. Once construction begins, it will not stop. [Stockbridge Tape 7, 1988:3]

5.2.3 Has the SSC been authorized, or does it have to go through Congress? [White Oak Tape 4, 1988:5]

The yearly appropriations to move the project on have been approved. But there is the distinct possibility that the SSC may never be built. [White Oak Tape 4, 1988:5]

5.2.4 At any time are there any major fees to update or maintain the SSC? [White Oak Tape 5, 1988:2]

Yes, new costs will evolve. For example, new magnets may be needed. [White Oak Tape 5, 1988:2]

6.0 HEALTH/SAFETY

6.1 RADIATION

6.1.1 What is the potential radiation hazard of the SSC? [Q Sheet 2/16/88]

The extent of radiation produced by the SSC is about equivalent to a university research facility or a modern hospital where radiation diagnosis and therapies are used. The radiation shielding and radioactive materials handling considerations for the SSC facility are similar to those that have proven to be effective at existing accelerator facilities. At the depth of the SSC tunnel at the Stockbridge site, a worst-case loss of beam occurrence would probably be undetectable anywhere the general public might be, including directly over the tunnel. An accelerator has no inherent source of radioactivity that could be released in an accident, and there is nothing to sustain a chain reaction. The source of energy is the same commercial power grid that serves a community's homes. [Gilbert/Commonwealth Inc. 1987; Universities Research Association nd; Jones 1988]

Everyday, everyone on earth is exposed to radiation. The SSC will produce some radiation as a by-product of particle collisions, but it will be absorbed by the detectors and by the soil covering the tunnel. An accelerator is something like a very large scale version of the projection tube in a television set. It produces a stream of particles that can be switched on and off like a light. It has no inherent source of radioactivity that could be released in an accident, and there is nothing to sustain a chain reaction. The source of energy is the same commercial power grid that serves a community's homes. [Universities Research Association nd]

6.1.2 Are people safe from radiation if they drive back and forth over the top of the tunnel on a daily basis? [Harshbarger 1987]

Passing over the tunnel is safe. To give some indication, the two most recent measurements of radioactivity in the air above the particle accelerator at Fermilab were far below federal limits. At one measurement, radiation was recorded at about one thousandth of the limit for one year, and no radiation was detected during the other. No radiation was detected in drinking water for either year. The Fermilab tunnel is 20 feet below the earth's surface and is safe. For geological reasons, the SSC in Michigan would be at least 100 feet below the surface.It will be perfectly safe to live permanently directly over the tunnel. [Jones 1988; SSC Central Design Group 1987:10]

6.1.3 But if it is safe to cross over the tunnel and live within its 53-mile circumference, why don't people live within the circumference of other accelerators? [Harshbarger 1987]

At Fermi, the accelerator ring is about 4 miles long. The amount of land it encircles is so small that the federal government purchased all of it and decided to use the space for park land rather than residences. The Super Proton Synchrotron that straddles the French-Swiss border near Geneva, Switzerland, at the CERN Laboratory is about the same size as the Fermilab accelerator. Houses and farms are around, over and within the accelerator ring, which is also tunneled underground. A new accelerator with about a four-mile circumference is nearing completion tunneled under the city of Hamburg, Germany. The largest accelerator under construction currently is LEP, also at the CERN laboratory. It is 17 miles in circumference and passes under several villages, as well as roads, farms, et cetera. Thousands of people live in the area encircled by that tunnel. For the SSC, the ring is much larger and there is too much land inside for the federal government to purchase all of it. Therefore, property owners in that area will continue their present land uses. [Jones 1988; Kane 1988a; Traugott 1988; Burdock 1988a]

6.1.4 Where in the ring is the radioactive material? [Q Sheet 2/16/88]

Most of the radioactivity will be in the beam absorbers (aborts). A minor amount may also be produced in beam collimators and scrapers at certain points within the tunnel and in the experimental detectors. [Jones 1988]

6.1.5 What happens if the cooling systems fail?) [Harshbarger 1987]

Nothing will happen that will pose a threat to the public. The machine will just be fixed. [Kane 1988a]

6.1.6 What effect can radioactive particles possibly have on individuals living in the abort zone? Is there any scientific data to support these effects or lack of effects? [Response Card 2/16/88]

None. No radiation detectable above normal background levels will be produced at the ground surface above the beam aborts. There is good evidence and data from 15 years' operations at Fermilab and European accelerators to support these designs, calculations and conclusions. [Jones 1988]

6.1.7 What happens to the proton beams that are ejected into beam abort dumps? What is the half-life of the radiation in the abort dumps? [Harshbarger 1987]

Proton beams will be directed to the beam absorbers (or aborts or dumps, as they are sometimes called) at the end of each accelerator cycle once or twice per day. Most of their energy is converted to heat. The materials used in the absorbers - carbon surrounded by aluminum - are chosen so that the small fraction of beam energy which goes into nuclear collisions results in few radioactive isotopes. The residual radioactivity in these absorbers will be dominated by 22Na and 60Co, with half-lives of a few years. [Jones 1988]

6.1.8 If severe cracks develop in the underground structure, how would you detect these cracks, and would any radioactive debris escape into the earth or ground water? [Harshbarger 1987]

If severe cracks develop in the beam absorber structures, ground water percolating through the crack could become slightly radioactive. For this reason ground water will be continuously monitored by laboratory personnel. There is no reason to believe that such failures would develop. [Jones 1988]

6.1.9 Explain the 650-foot radiation exclusion zone and the necessity for it. [Harshbarger 1987]

The 650-foot radiation exclusion zone is a zone in the plane of the ring (horizontally, at the depth of the ring) extending radially outward that distance (but vertically only about 20 feet high) to attenuate the most penetrating radiation, particles called muons, which would be produced in a worst-case beam loss at any point on the machine circumference. Muons produce ionization and would be an immediate radiation source in such an event, but they do not produce residual radioactivity and thus would not contaminate soil or ground water. [Jones 1988]

6.1.10 What will be the half life of the radiation on the ground above the tunnel? [Harshbarger 1987]

There will be no measurable radiation due to the SSC on the ground surface above the tunnel and no radioactivity will be produced there. [Jones 1988]

6.1.11 If radiation is to be escaping around the tunnel, how will it be patrolled to keep people from making contact? How safe are people going to be who live near the collider? Who is going to pay for the cost of patrols and security? [Harshbarger 1987]

In spite of a design to prevent any radiation from escaping around the tunnel, the SSC Laboratory will have a safety officer and radiation safety professional personnel who will be responsible for continuously monitoring air, water and materials (soils, machine components, etc.) for radioactivity and for other noxious materials (e.g. ozone). Areas where an accident would result in radiation exposure will be interlocked to preclude entry during laboratory operations. The cost of radiation safety monitoring, interlocks and staff will be borne by the SSC Laboratory. It will be entirely safe to live near or directly above the Collider. [Jones 1988]

6.1.12 Will the soil and water be contaminated in future years by the buildup of radiation and toxic waste?

The soil and water will not be contaminated in future years by a buildup of radiation or of toxic waste. Radioactivity will be confined to removable components, such as the beam absorbers. The toxic waste generated will be mostly cleaning solvents, etc. and will be handled in an environmentally appropriate manner, as at a modern university or industrial laboratory. [Jones 1988]

6.1.13 What will happen if an intruder animal - a squirrel, cat, etc.- were to get into the tunnel? [public meetings...]

Under normal operating conditions, the animal would probably not experience a significant radiation exposure, but it might be electrocuted. [Jones 1988]

6.2 WASTE

6.2.1 What types of waste, and how much of each, will be generated by the SSC? [Hot Line 1988a]

In order to determine the types and amounts of waste that the SSC will generate, the Department of Energy has reviewed waste generation data from other basic research facilities. The types of waste considered can be broken down into these categories: conventional, hazardous, radioactive and mixed (radioactive hazardous waste). To put the estimated amount of waste from the SSC in perspective, the following comparisons can be made:

1. The amount of conventional waste is equivalent to a light industrial facility with a similar number of employees and visitors. An estimated 30,000 cubic yards/year of solid waste, and 150,000 gallons/day of liquid waste (sewage) will be generated.

2. The amount of low-level radioactive waste is about the same as that from a major university with a medical facility. An estimated 300 cubic yards/year will be generated, although the amount may be reduced to 15 cubic yards/year. Experience at Fermilab has shown that 95 percent of the material disposed of as low level radioactive waste shows no evidence of radioactivity. If such wastes are screened, a 95 percent reduction in volume may be realized, and a similar situation is probable at the SSC.

3. The hazardous waste generated is like that from a small industrial facility, and is estimated to be about 10,000 gallons/year.

4. Mixed waste is estimated to be less than 0.1 cubic yards/year, with the probability that it will be zero. [United States Department of Energy 1987c; Jones 1988]

6.2.2 If this is built in Michigan does this mean that toxic waste will be brought in from other states and/or counties to be disposed of? Is the SSC an inducement for a radiation disposal site in our area. [Harshbarger 1987]

The waste produced from the SSC, radioactive or otherwise, would not be an incentive to place additional toxic or radioactive disposal sites in the state. Quantities of either type of waste produced by the SSC are not economically sufficient to support the need for additional disposal sites. [Heinzman 1988c]

Toxic waste is a by-product of many industrial activities, and there is no connection between the SSC and the location of toxic waste sites. In addition, the SSC is not an inducement for a radioactive disposal site, something which is determined by other criteria. [Kane 1988a; Burdock 1988a]

6.2.3 How and where are toxic and radioactive wastes to be disposed of? [Harshbarger 1987]

Toxic wastes will be handled under existing State of Michigan laws for the handling of toxic and hazardous materials. The federal and state requirements for the final disposal of these materials will be followed. The disposal of radioactive waste from this facility will be handled through existing DOE low-level radioactive disposal sites outside of the State of Michigan. The legislative act governing the disposition of low-level radioactive waste in Michigan states that SSC radioactive waste will not be placed in our future low-level waste repository. [Heinzman 1988c]

Radioactive wastes would be sent to a DOE low level waste repository, such as Hanford, Washington. Toxic waste (cleaning agents, etc.) would be incinerated or disposed of as at any other modern university or industrial laboratory. [Jones 1988]

6.3 WATER AND RADIATION

6.3.1 Will radioactivity contaminate the ground water? [White Oak Tape 4, 1988:3]

The SSC was conservatively designed to prevent any such contamination from occurring. Data from each of the seven competing states has been collected by the Department of Energy specifically to address the potential impacts of the SSC on groundwater. The DOE must acknowledge this possibility in selecting the final site. If a site poses major problems, the DOE will not locate the SSC in that state. [Heinzman 1988a, 1988c]

6.3.2 Would you alert the people if there were a spot with warm water? [Q Sheet 2/16/88]

A monitoring program will be constructed that is capable of detecting minute changes in the existing radiation found in groundwater. The public would be informed of such a problem in the case that it ever did happen. [Heinzman 1988c]

6.3.3 How often will wells be monitored throughout the construction and operation of the project? [Q Sheet 2/16/88; Response Card 2/16/88]

During construction, wells will be monitored when there is a question of affecting the water table. The frequency will depend on construction details, unknown at this time. During operation, special wells will be used to monitor for evidence of radioactivity near beam absorbers. The frequency will depend on water percolation rates and other engineering factors still undetermined. [Jones 1988]

Any groundwater program that may be required for this project will meet state and federal guidelines. The frequency of monitoring depends upon the nature of introduced contaminants into an aquifer system. There is no reason to expect the tunneling method of construction will introduce contamination into the aquifer. After the SSC has begun operation, a formal environmental monitoring plan will be designed for the SSC site facilities. [Heinzman 1988c]

6.4 MISCELLANEOUS SAFETY

6.4.1 Will the facility be a target in the event of a nuclear attack? [Q Sheet 2/16/88; White Oak Tape 5, 1988:2]

It is highly unlikely that the SSC would be a target since it is used purely for research and is engaged in the pursuit of knowledge. [White Oak Tape 5, 1988:2]

6.4.2 Does the SSC produce any magnetic fields that are dangerous to people or animals? [Stoffle et al, 2/16/88]

The SSC does not produce magnetic fields detectable at the surface of the ground or even close to the accelerator tunnel. Magnetic fields from the SSC at any location where people or animals might venture are small compared to the Earth's magnetic field. [Jones 1988]

7.0 QUALITY OF LIFE

7.1 COMMUNITY VALUES

7.1.1 What will happen to our community rural values if the SSC is built here? [Harshbarger 1987]

7.1.2 How will the nature of small communities change if the SSC is built here? [Harshbarger 1987]

7.1.3 What will be the relationship between the people of Stockbridge and those of the incoming campus community? [Stockbridge Tape 7, 1988:2]

It is difficult to predict where the SSC people would locate, and therefore, what their relationship to the community would be. One possibility was mentioned by Vivian Lund, mayor of Warrenville, Illinois, a neighboring community of the Fermi National Accelerator Laboratory. Mayor Lund explained that enhancement of science education in schools near Fermilab had improved. "High school science classes visit the lab, and Fermi people visit area schools," she said. "The availability at all levels has improved." [Town Crier 3/1/88]

7.2 GROWTH

7.2.1 Will a boom town result from the SSC? [Harshbarger 1987]

7.3 MISCELLANEOUS

7.3.1 Does the SSC present any potential restriction of airspace? Any restriction of aircraft over flights? [Response Card 2/16/88]

The SSC presents absolutely no restriction of airspace or overflights. [Jones 1988]

7.3.2 Does the SSC interfere with television reception? [Hot Line 1988b]

No television interference will occur because of the SSC. [Hot Line 1988b]

7.4 RELOCATION

7.4.1 How can SSC officials justify relocating farm families that have lived together in extended family networks for years and years? [Harshbarger 1987]

7.4.2 How can SSC officials justify gaining ownership of a centennial home? [Harshbarger 1987]

8.0 REAL ESTATE

8.1 ACQUISITION PROCESS

8.1.1 What is the timetable for buying land? [Harshbarger 1987]

REAL ESTATE ACQUISITION SCHEDULE

SPRING 1988-PRE-APPRAISAL OF PROPERTIES
-This phase includes taking aerial photographs of proposed areas in Ingham and Jackson counties.
-Title work will be prepared to determine ownership of properties.

SUMMER 1988-APPRAISAL OF PROPERTIES
-Research will be done to determine market values of underground property need.
-Appraisals begin.

PRE-ACQUISITION OF PROPERTIES
-Procedures for acquisition and relocation will be developed with special emphasis on properties in the campus, future expansion and service areas and the near cluster. (These properties are expected to be the first transferred to the DOE in June 1989).
-A relocation plan will be completed that will be used to determine individual needs of each affected person/family.
-An inventory of available house and an individual analysis of each displaced person's needs will be completed.
-Meetings will begin to explain relocation program, plus advisory services to assist people who will move.

FALL 1988-NEGOTIATIONS
-Negotiations will begin with owners of properties to be acquired in first half of 1989 to obtain options or to purchase properties.

1989-ACQUISITION BEGINS
-June 1: Campus, first 25 percent of collider properties transferred to the DOE.
-Fall: Property Management: Security and maintenance of all purchased properties will be provided. Removal of any acquired improvements will begin.

-September 1: Injector, second 25 percent of collider properties transferred.
-December 1: Third 25 percent of collider properties transferred.
-March 1, 1990: All remaining areas transferred.
[Michigan SSC Study Team 1988]

8.1.2 When will property owners be notified about selling property? [Q Sheet 2/16/88]

Property owners will be notified in the summer of 1988. [Hohl 1988]

8.1.3 When will appraisals and options be undertaken? [Hot Line 1988b]

The process is scheduled for late summer 1988. The contracts are still being worked on. [Hot Line 1988b]

8.1.4 When will payments begin? [Q Sheet 2/16/88]

Payments will begin as soon as possible after Michigan is selected, if the Stockbridge, Michigan, site is chosen. [Hohl 1988]

8.1.5 How will land be acquired? [Q Sheet 2/16/88; Hot Line 1988a]

Through negotiations with affected owners, the compensation due the owners will be based on fair market value appraisals. In situations where only a portion of an owner's property is needed for the SSC project, the compensation due the owner will be based on the difference between the value of the property as it presently exists and the value of the remaining property after the taking of the land needed for the project. [Hohl 1988]

8.1.6 Who is going to acquire the land? [Hot Line 1988a]

The State of Michigan will purchase and acquire the land and will then deed ownership to the Federal Government. [Jones 1988] (Gilbert/Commonwealth Inc., a qualified contractor in Jackson, Michigan, will acquire the land under contract with the State of Michigan. [Hohl 1988])

8.1.7 Who will own the land? [Hot Line 1988a]

The State of Michigan will purchase and acquire the land and will then deed ownership to the Federal Government, specifically to the Department of Energy. [Jones 1988; Hohl 1988]

8.1.8 Will the land be leased back to the farmers? At what cost? [Hot Line 1988a]

If the Department of Energy permits it, land will be leased back to the farmers at fair market value. [Hohl 1988]

8.1.9 Will land use or homes within the collider ring boundaries be affected in any way? [Harshbarger 1987]

These lands are not part of the project and will not be affected in any way. [Gilbert/Commonwealth Inc. 1987]

8.1.10 How many homes will be affected by the SSC project? [White Oak Tape 5, 1988:2]

Because of the large land requirements of the SSC, a number of homes and farms will have to be purchased by the State. Between 200 and 225 homes will be affected at the Stockbridge site if it is selected. [Gilbert/Commonwealth Inc. 1987; Burdock 1988a]

8.1.11 If the government decides that it does not want the SSC after it has already bought the land, what will happen? [Q Sheet 2/16/88]

The first right to purchase the land will go to the previous owner. [Q Sheet 2/16/88]

8.1.12 Will the government pay to survey? [Q Sheet 2/16/88]

The government would not pay to survey since it has the right to do so. Any damage resulting from this process would be compensated. [Q Sheet 2/16/88]

8.1.13 Will the surface property be rezoned? [Q Sheet 2/16/88]

No, there is no reason to rezone the property [Q Sheet 2/16/88]

8.1.14 When will land for the future expansion area be acquired? [Q Sheet 2/16/88]

The land for that would be purchased now. [Q Sheet 2/16/88]

8.1.15 Does the government buy land not accessible to the farmer whose land is cut off by the SSC? [Q Sheet 2/16/88]

The government will buy land to which the landowner has no access. [Q Sheet 2/16/88]

8.1.16 How will individuals with land currently enrolled in Public Act 116 be affected by sale of their property to the state for the SSC project? [Harshbarger 1987]

The Michigan DNR has stated that landowners whose property is enrolled in Public Act 116 and is required for the SSC project will not be required to pay any penalties. [Gilbert/Commonwealth Inc. 1987]

8.2 DEED RESTRICTIONS

8.2.1 What are the deed restrictions on the lands of stratified fee estate purchases? [Hot Line 1988a]

Two restrictions required by the Department of Energy will be to 1) limit activity that causes vibration, and 2) restrict activity that would penetrate the collider ring. [Hohl 1988]

8.2.2 How can I sell a piece of land that I have leased to an oil company? [Hot Line 1988a]

Land will be acquired subject to the lease or acquisition, which will include the oil company as a party of interest. [Hohl 1988]

8.3 FEE SIMPLE AREAS

8.3.1 What is "fee simple?" [Hot Line 1988a]

Fee simple involves people with land/homes on the SSC ring who will be asked to sell their land, fee-simple-title, to the state. [Stoffle et al., 1988]

8.3.2 Can you cross fee simple property? [Q Sheet 2/16/88]

Except where there are buildings, which may have some restrictions, people will be able to cross fee simple property. There will not be a fence constructed around the whole SSC facility and people are invited to visit the site. [Grover 1988]

8.4 MINERAL RIGHTS

In most cases, because of drilling restrictions to protect the integrity of the collider tunnel, the State will purchase mineral rights from the landowner. The value of the gas and oil resources will be determined in the appraisal process. However, it is expected that the appraisal of mineral rights will be of concern only in a few of the properties.

8.4.1 Do I get to keep the mineral rights in a fee simple area? [Hot Line 1988a]

According to the DOE, the State has three possibilities to consider: (1) The original owner would retain mineral rights; no development would be allowed within a safety zone on either side of the tunnel ring, (2) The State could purchase the mineral rights outright and retain them, or (3) purchase the rights and turn them over to the DOE for its control. In June 1988, the State will propose which possibility to pursue for the fee simple areas. [Heinzman 1988b]

We are requesting a clarification of this question from the Department of Energy. [Hohl 1988]

8.4.2 Do I get to keep the mineral rights under a stratified fee area? [Hot Line 1988a; White Oak Tape 4, 1988:1]

According to the DOE, the State has three possibilities to consider: (1) The original owner would retain mineral rights; no development would be allowed within a safety zone on either side of the tunnel ring, (2) The State could purchase the mineral rights outright and retain them, or (3) purchase the rights and turn them over to the DOE for its control. In June 1988, the State will propose which possibility to pursue for the stratified fee areas. [Heinzman 1988b]

We are requesting a clarification of this question from the Department of Energy. [Hohl 1988]

8.5 OPTIONS

8.5.1 Is the State of Michigan going to pay for an option? [Q Sheet 2/16/88]

The State is developing a compensation plan for landowners who will be required to sell their land to the State for the SSC. Part of this plan includes an option payment (see question 8.5.5).

8.5.2 What is the State schedule on the option to build? [White Oak Tape 4, 1988:5]

Notices will be sent to property owners before surveyors go in. [White Oak Tape 4, 1988:5]

The DOE has made available a schedule for the transfer of title of property targeted for the SSC (see question 8.1.1). This property acquisition schedule specifies that one-fourth of the property will be transferred to the DOE at a time. Before the DOE announces the final SSC site late this year, the State will work with property owners who live on the first quarter of property to be transferred. The State and property owners must reach agreement on the purchase price of property and on the State's exclusive right to purchase the property.

8.5.3 Where can people get option papers? [Q Sheet 2/16/88]
Undecided at this time. [Hohl 1988]

8.5.4 What will hold options? Word of agreement or a percentage down? [Q Sheet 2/16/88]
Undecided at this time. [Hohl 1988]

8.5.5 What percentage of the purchase price will the option be? [Hot Line 1988b]
The State is in the process of creating an inducement to the option to enhance the likelihood that it will be satisfactory to property owners. All property owners who sell land to the State for the SSC will receive five percent of its appraised value, with a minimum payment of $500, if the owner signs the option agreement within 30 days of the original offer. The option payment will supplement the purchase price of the property. Option payments will remain in effect throughout the entire land acquisition program in fairness to property owners who will be approached after DOE has selected the Stockbridge site. The options program is not final, but the State is working for its acceptance.

8.6 PROPERTY VALUES

8.6.1 What is the anticipated impact of the SSC on property values? [Harshbarger 1987]
Property values should increase for residential and commercial areas that may provide homes for SSC staff or commercial establishments that directly or indirectly provide services to the SSC. Additionally, property values for areas that are considered ideal locations for homes or future high technology development will

increase. Other areas should be unaffected. [Gilbert/Commonwealth Inc. 1987]

8.6.2 What has happened in Illinois, and what may happen here to property values inside and around the oval? [Response Card 2/16/88]

Batavia, Illinois, the host of Fermilab, has experienced a slow and steady increase in property values.

In Illinois, land inside the Fermilab ring was acquired by the Department of Energy and is used as open space. This will not occur in Michigan (see 6.1.3). [Hohl 1988]

8.6.3 Will property value inside the SSC ring be affected? [Q Sheet 2/16/88] Only the property near the campus area should be affected. The value of this property probably will go up. [Q Sheet 2/16/88]

8.6.4 Many questions about specific areas on the map, and the effect on property values in these areas. For example, "I live within the 'I' area and across from the 'J' area. What effect will these areas have on my property value? Will any of my land need to be purchased. [Q Sheet 2/16/88; Response Card 2/16/88]

It is improbable that the J areas will be developed for at least a decade, and the underground use of the I areas will also be well into the future. The DOE wishes to acquire them now to have them available, although no budget for their development is in the SSC budget as presented to the DOE. (see 2.3.5 and 2.3.6) [Jones 1988] A project office will be established at the site, where information of this nature can be gained. [Hohl 1988]

8.6.5 What effect will the abort area have on land value? [Hot Line 1988b]

The abort area will not have any special effect on the land value because there is no effect on the surface. [Hot Line 1988b]

8.7 RELOCATION & COMPENSATION

8.7.1 How will landowners and homeowners be compensated for the loss of their property? [Harshbarger 1987]

The State of Michigan has prepared a land acquisition plan that will be included in the proposal submitted to the Department of Energy. This plan outlines procedures for appraisal, compensation, and condemnation proceedings, if required. Property owners will receive current full market value and relocation assistance. Other programs are still being developed. [Hohl 1988; Gilbert/Commonwealth Inc. 1987]

In March 1988, the Michigan SSC Study Team prepared a written statement that outlined four components of the land acquisition program to be offered by the State. This statement was prepared to reduce confusion in the Stockbridge area, which resulted from the dissemination of inconsistent information about the land acquisition program. The information outlined in this statement is summarized below. [Michigan SSC Study Team 1988]

Land Acquisition Responsibilities

The Michigan Department of Transportation (MDOT) has served as an advisor to the SSC Commission on the land acquisition program. MDOT was selected because of its familiarity with state and federal land acquisition and relocation laws. The firm of Gilbert/Commonwealth Inc., of Jackson, Michigan, has been awarded the contract to manage the land acquisition and relocation program, under the supervision of the SSC Commission.

The Four Parts of Land Acquisition

Payment of Fair Market Value for Property: The basic component of the land acquisition is a purchase price for property based on fair market value. Supplemental compensation programs will be described below. The fair market value for property will be determined by professional appraisers working under contract for Gilbert/Commonwealth Inc. The appraisals will reflect the property value at its highest and best use.

Equity Adjustment Payments for Agricultural Lands: The Michigan state constitution requires the State to purchase land for fair market value. A proposal is being prepared that provides for equity payments to agricultural landowners as compensation for possible declines in farmland values that may have occurred in the last 10 years. This payment would supplement the fair market purchase price and any relocation assistance payments for which they may be eligible. Under the direction of the Michigan State University Cooperative Extension

Service, a survey of land prices in the targeted areas of Jackson and Ingham counties is being conducted. If agricultural land values are found to have declined, legislation will be proposed to give farmland owners an equity payment to compensate for any lower, current prices. The Cooperative Extension study team will make its recommendation to the SSC Commission before the summer of 1988. Its findings will be used to formulate appropriate legislation to enact the program.

Relocation Assistance Programs: In addition to purchasing property at fair market value and equity payments for agricultural lands, the State will pay residents the costs of relocation. The Uniform Relocation Assistance and Real Property Acquisition Policies Act of 1970 (Public Law 91-646) applies to people who must relocate should the SSC be located in Michigan. The law provides for a wide variety of relocation benefits, including:

* Assistance in finding a suitable replacement property.

* Payment of moving costs, either the actual, reasonable cost of moving, or payments based on a fixed schedule. The relocating person chooses which payment is most suitable. When necessary, moving costs will also cover related expenses, such as transportation to the new home, payments for temporary housing, and the storage costs of personal property.

* Housing payments are separated into three types:(1) Payment for replacement housing, (2) Rental supplements, and (3) Downpayment for a new home for displaced tenants.

* Interest differential payments to compensate for an increase in interest costs on mortgages. To receive full reimbursement, those applying for relocation benefits must have proper documentation of expenses.

Other Assistance Programs: The SSC Commission will provide an ombudsman to help relocated individuals and families take advantage of a wide variety of existing county, state and federal programs. These programs can help ease adjustments to relocation. The programs include the federal Soil Conservation District tree planting projects, the Michigan Department of Natural Resources wildlife stocking and habitat development programs, and stress counselling for displaced people. In addition, the Cooperative Extension Service has a number of programs to help relocated farmers plan, manage and improve new farms, fields and herds.

8.7.2 Landowners will be paid for their land; but what about landowners who rent out their land? How will they be compensated for this income? [Stoffle et al., 1988]

Assuming the rent is fair market rental, the owner would receive compensation sufficient to acquire comparable replacement land that should produce the same rental as the land acquired for the construction of the SSC. Another possibility is that the compensation received could be put into other investments that could return an income equal to the lost rental. [Hohl 1988]

8.7.3 Will people that have their house taken be given any help for relocation? [Harshbarger 1987]

Yes, everyone displaced because of the SSC will be entitled to an eligibility determination and assistance under the Federal Uniform Relocation Assistance and Real Property Acquisition Policies Act of 1970. [Hohl 1988]

8.7.4 How can the elderly cope with moving out of their homes? Will special programs be developed to meet the needs of the elderly persons who relocate? [Harshbarger 1987]

Special needs assistance will be provided in accordance with provisions under the Uniform Act mentioned in 8.7.3. [Hohl 1988]

8.7.5 What are the different types of assistance? What are the amounts in dollars? [Hot Line 1988a]

The following types of assistance are available:

* Advisory assistance to help with problem solving.
* Real estate referrals to assist in finding a new location.
* Financial grants to assist in relocation. These grants cover moving personal property, and housing and rental costs for residential occupants. [Hohl 1988]

The amounts in dollars is listed below:

* Residential moving costs based on actual receipted bills from a moving company or a schedule based on room count of up to $500.00.
* Housing grants for long-term owner occupants of up to $15,000.00

* Rental and purchase down payments for tenants and short-term owner occupants of up to $4,000.00.
* Business moving costs based on actual receipted bills from a moving company or a lump sum amount or a payment in lieu of moving based on business net income for one year not to exceed $10,000.00. [Hohl 1988]

8.7.6 Can houses be moved from the purchase site under the relocation program? [Hot Line 1988a]

House retention at salvage value is permitted during negotiations. The costs to move and replace the house would be the relocatees' cost. Relocation funds would be available under limited conditions. [Hohl 1988]

8.7.7 Will houses above the SSC ring be removed? [Q Sheet 2/16/88]

On stratified fee estate areas, no.

8.8 STRATIFIED FEE AREAS

8.8.1 What is "stratified fee estate?" [Hot Line 1988a; Q Sheet 2/16/88]

Stratified fee estate involves people with land/homes on the SSC ring who will be asked to sell a lower strata of ground to the state for the ring at about 140 feet below the surface. [Stoffle, et. al, 1988]

8.8.2 How is the value of a stratified fee purchase determined? [Hot Line 1988a]

The value of a stratified fee purchase will be determined by the difference between the Before Value of the property unaffected by the SSC, and the After Value of the property encumbered by the stratified fee acquisition. [Hohl 1988]

8.8.3 What is the basis or formula for determining the amount of money to be paid home-, landowners residing in the stratified fee estate areas? [Response Card 2/16/88]

A comprehensive market study of property affected by similar stratified fee acquisitions will be made to determine the compensation due property owners in the subject stratified fee estate areas. The findings of this study will be incorporated in the after value appraisal mentioned in 8.8.2. [Hohl 1988]

8.8.4 Is the fact that property values are likely to fall in this area taken into account when this fee is paid? [Response Card 2/16/88]

Property values are not expected to fall in the stratified fee area. Land values in the stratified fee area will equal the sum of parts of the total value of the fee simple estate. (Value of land of owners' estate + value of the Department of Energy's estate = value of fee simple estate.) [Hohl 1988]

8.8.5 Can people build on top of the stratified fee estate areas? What restrictions are there? [Q Sheet 2/16/88]

Yes, building will be allowed on top of the stratified fee areas. Two known restrictions will limit activity causing vibration and penetration of the collider ring. (see 8.2.1) [Hohl 1988]

8.8.6 What does the state plan to do about the devaluation of the land bought up in the stratified fee estate? [Response Card 2/16/88]

A loss of value for this land is not expected. Compensation for deed restrictions will be determined by a comprehensive market study. [Hohl 1988]

9.0 ROADS

9.1 CONSTRUCTION/MAINTENANCE

9.1.1 Will any roads be closed due to construction or operation of the SSC? [Hot Line 1988a]

It is too early to know for certain if specific roads will be closed for a short time. However, SSC officials do not anticipate any substantial problems with road closures. If the need arises to close a road temporarily, alternative routes would be made available. No long-term road closures are intended. [Ott 1988a]

9.1.2 Will there be any new roads? [Q Sheet 2/16/88]

Studies are being done by the Michigan Department of Transportation about road usage. Present roads will be used when possible.

9.1.3 Will highway M-52 be widened to four lanes, and if so, when and where? [Hot Line 1988b]

There are no plans to widen M-52. [Hot Line 1988b]

9.1.4 Would county roads be upgraded? Who would pay for this? Who would pay for repairs? Who pays for new roads? [Harshbarger 1987]

9.1.5 Who will bring the roads up to acceptable standards around construction and the ring? [Harshbarger 1987]

9.1.6 What units of government would be responsible for upgrading and maintaining the improved roadways used by the SSC project? [Harshbarger 1987]

The State of Michigan has assembled a package of incentives for the SSC project that includes improvements to the existing highway and road network needed to support the construction and operation of the SSC. It is not anticipated that any local unit of government will have to spend any funds beyond those which are currently being expended for road maintenance and construction. [Gilbert/Commonwealth Inc. 1987]

9.1.7 Where is the money coming from for road expansions? [Stockbridge Tape 7, 1988:1]

State funds will be used for project-related expansion and maintenance. [Stockbridge Tape 7, 1988:1]

9.2 TRAFFIC

9.2.1 Will the SSC increase local traffic? [Harshbarger 1987]

While the SSC will increase traffic during construction and operation, access routes will be selected which minimize the impact on residential and commercial areas. Heavily travelled routes will be upgraded to accommodate the increased traffic. [Gilbert/Commonwealth Inc. 1987]

9.2.2 Can the road commission handle extra work, traffic and congestion? [Harshbarger 1987]

10.0 SERVICES

10.1 ELECTRICITY

10.1.1 How much electrical power is needed? [Stockbridge Tape 8, 1988:2]

The refrigerators used to cool the magnet system will require about 30 megawatts of electric power. This amount is comparable to or slightly less than that being used to operate the largest existing accelerators. (Gilbert/Commonwealth Inc., 1986:7,8)

The accelerator runs 24 hours per day, seven days per week. The power consumption is about 50 megawatts to 100 megawatts... [Stockbridge Tape 8, 1988:2]

10.1.2 Where will the extra electrical power come from? We now have Detroit Edison and they constantly are raising their costs. Their reason is they do not have enough power or energy. Is this going to raise our costs? [Response Card 2/16/88]

The electric power proposed for the SSC will come from the Michigan Electric Coordinated Systems (MECS), which comprises the electric generating and transmission systems of Consumers Power Company and Detroit Edison Company. CPCo and DECo anticipate no problem in reliably and economically serving the SSC. Such a small increase in energy generation would not cause any significant increase in fuel costs. [State of Michigan 1987:8-1,8-17]

10.1.3 Will Consumers Power provide electricity? [White Oak Tape 4, 1988:5]

Up to 200,000 kilowatts (200 Megawatts) are required. Consumers Power currently has this capacity in its system. [Jones 1988]

10.1.4 I am concerned about how the electricity needed for the SSC will be delivered. I have read in news reports there are concerns about the health risks of people living in close vicinity of high power tension electric lines. [Response Card 2/16/88]

The power drawn by the SSC complex, up to 200 Megawatts, is small compared to the power currently carried by the high tension lines crossing the site. [Jones 1988]

10.2 EMERGENCY SERVICES

10.2.1 What about emergency services for the campus and construction workers? Fire? Ambulance? The local volunteers will be hard pressed to handle a drastic increase in volume of runs. [Response Card 2/16/88; Hot Line 1988a] The SSC will most likely maintain its own fire and security protection. The current design of the SSC facilities calls for a small building in the Central Campus area that would house fire, rescue and site-control operations. The building will be located on one of the main site roads to ensure easy and rapid access of emergency vehicles. [SSC Central Design Group 1986:553]

10.2.2 Will the state and federal government pay for the extra fire and police protection? Will they subsidize the local communities? [Response Card 2/16/88]

The SSC most likely will maintain and pay for its own fire and security protection. In the case of Fermilab, there are mutual assistance compacts between the laboratory and neighboring communities. [Jones 1988]

10.3 SEWAGE

10.3.1 What will be done to accommodate additional sewage created by influx of new residents and workers at the SSC?) [Response Card 2/16/88; White Oak Tape 4, 1988:4]

The city of Stockbridge sewage treatment plant will handle the additional waste. It will be necessary to upgrade the plant to accommodate the additional loading. [Heinzman 1988c]

The sanitary network for the campus, including the injector area, will be served by a pipe collection system discharging to an existing or planned sewage treatment plant. The sewage plant effluent discharge is estimated at 150,000 gallons per day. [State of Michigan 1987:8-21]

10.3.2 What about sewage treatment for expanded county development? [Harshbarger 1987]

The need for planning assistance and community development programs to accommodate growth at the county level are acknowledged as part of the State of Michigan proposal. As indicated in the proposal, assistance will be provided to communities in need of planning advice. [Heinzman 1988c]

10.4 WATER USE OF SSC

10.4.1 How much water is required by the SSC, and where will it come from? [Response Card 2/16/88]

Water availability is a consideration in evaluating site suitability. A potential need for up to 2,200 gallons per minute (peak) of industrial water exists. In addition, 250 gallons per minute (average) of potable water is required. Appropriate waste water disposal facilities are necessary for a laboratory of 3,000 people. [Appendix C, SSC Siting Considerations; Section C-6, Utilities]

The SSC will require approximately 2,400 gallons of water per minute. New water sources are currently being identified so there will be no effect on the water supplies of current users in either of the project areas. [Gilbert/Commonwealth Inc. 1987]

10.4.2 Will any water towers be constructed to service the facility? Where? [Hot Line 1988a]

It may be necessary to construct a water tower to regulate the water pressure for the central campus area. The possible location of the tower, if necessary, is not known at this time. [Heinzman 1988c]

10.4.3 What will water be used for? [Hot Line 1988a; Harshbarger 1987]

Drinking and sanitary water will be necessary for the 3,000 employees at the central campus area. Additional water will be needed to run the cooling towers at selected points around the ring. [Heinzman 1988c]

10.4.4 Is water used for the cooling system? If so, how? How much water will be used? [Hot Line 1988a]

Water will be used for part of the cooling system. The use will be through cooling towers that will remove waste heat from the refrigeration units. The industrial water use was not broken down into an exact amount for cooling; however, it can be expected to be less than the 2,200 gallons per minute as indicated for a peak demand by the DOE for non-domestic water use. [Heinzman 1988c]

Cooling water will require up to 2,200 gallons per minute, mostly as make-up water to replace evaporation losses. The total volume of water in the cooling system is much greater than 2,200 gallons. [Jones 1988]

10.4.5 What will happen to water used for cooling? [Q Sheet 2/16/88] (Is water used for cooling?)

There are two cooling systems in use: cooling ponds and cooling towers. With cooling ponds, the water heated by the equipment is cooled in open ponds by evaporation. In the other, cooling "towers" (structures that might be as high as a two-story building) are used. Towers are preferable where space is at a premium. Ponds can be blended into the local landscape. In both cases, water must be continually supplied to make up evaporation losses. [Jones 1988]

Some of the water will be lost to evaporation through the cooling towers. The water that condenses can be recycled through large holding ponds that can provide habitat for water fowl. [Heinzman 1988c]

10.4.6 What is meant by "dewatering?" How will it take place? [Hot Line 1988a]

Dewatering does not mean that all of the water will be removed from the area and never returned. During construction, a temporary relief of water pressure may be needed at selected locations. This temporary relief is accomplished through pumping water wells at a rate sufficient to diminish the amount of water entering an excavation. When the excavation is completed, the wells are shut down and water levels in the ground will return to their previous position. There will be only temporary effects on the water table in an area, with no permanent impacts on water levels. If a residential water well is affected on a temporary basis during construction, an interim water supply will be provided to the resident by the State of Michigan. [Heinzman 1988c]

11.0 TAXES

11.1 EFFECT ON TAXES

11.1.1 Will the SSC affect taxes? [Stockbridge Tape 7, 1988:4; Hot Line 1988a; Q Sheet 2/16/88]

The SSC should stimulate steady and controlled growth that would gradually increase the tax base in surrounding communities and eventually benefit the tax payers. The State will pay local townships for the loss in tax base due to large parcels of land being turned over to the Department of Energy for the SSC. The State's financial compensation for lost tax base will continue until the growth resulting from the SSC contributes enough to the local tax base to compensate for that loss.

11.1.2 It looks like there will be a substantial loss of revenues from taxes because the government will own what is now tax producing property. Will the township be compensated for this loss? [White Oak Tape 5, 1988:2]

The State aims to reimburse any local units that would lose revenues as a result of the SSC, although exact formulas are not yet known (see question 11.1.1). A bill currently in the senate (sponsored by senator Nick Smith) addresses this issue. [Burdock 1988c]

11.2 TAXES AS A RESULT OF SELLING PROPERTY TO THE STATE

The State has hired the international accounting firm of Peat Marwick Main & Co. to prepare a brochure on the tax implications of selling property to the State for those people who will be required to do so. The State recognizes the complexity of this issue and has formed a committee of tax experts from Peat Marwick to make available this information sometime this summer.

11.2.1 If a landowner sells property to the SSC project, does the landowner pay any taxes on the sale of the property? [Hot Line 1988b]

The landowner will pay taxes, but the state is trying to minimize the tax burden. [Hot Line 1988b]

11.2.2 Is option money taxable? [Grobe 1988]

11.2.3 Are expenses for relocating a building tax deductible? [Grobe 1988]

11.2.4 Can a person receive payment from the State for land and home over an extended period of time, to spread out taxes? [Grobe 1988]

12.0 WATER

12.1 EFFECT OF SSC ON COMMUNITY WATER SUPPLY

12.1.1 Will construction of the SSC facilities affect water levels or quality? [Q Sheet 2/16/88; White Oak Tape 4, 1988:2]

The construction of the SSC will have no effect on water quality, and water levels will be unaffected as well. During construction of the underground components of the SSC, some dewatering may be required. The flow from the dewatering will be diverted to existing surface drainages according to guidelines established by local and state permitting authorities. Any impact on local well yields will be mitigated by the state government. [Gilbert/Commonwealth Inc. 1987]

12.1.2 Will construction of the SSC project interrupt water flow, and if so, for how long? [White Oak Tape 4, 1988:2]

There is a commitment on the part of the state to anticipate any water flow interruptions and to provide alternative water sources of the same quantity and quality during any short-term interruption. No long-term interruptions are expected. [Stoffle; Burdock 1988a; Heinzman 1988a]

12.1.3 Will there be any damage to the wells? [Q Sheet 2/16/88]

If so, they would be replaced. People who have to relocate their wells will be compensated. [Q Sheet 2/16/88]

12.1.4 Who absorbs the cost of water lines or wells that are disturbed? [Harshbarger 1987]

The State of Michigan will provide water to residents temporarily influenced by construction activities for the SSC. [Heinzman 1988c]

12.1.5 How are you going to guarantee our water? [Stockbridge Tape 7, 1988:1]

The water resources will be protected by state and federal laws, which do not allow the impairment of groundwater and surface water. All state and federal laws and permits needed for construction will be obtained. This is not an industry that has a need for large volumes of water to treat or dilute contaminants. The only effects on water resources may occur during construction, and they would be associated primarily with the potential for dewatering at select locations (see 10.4.6). [Heinzman 1988c]

12.1.6 How is the disruption of existing wells (deep) to be affected in terms of drawdown or contamination? [Response Card 2/16/88]

The SSC is not a manufacturing industry that stores large volumes of raw or waste products in need of processing. Water use is primarily for cooling units or domestic application. Waste water from the domestic use will be handled by the Village of Stockbridge waste water treatment plant. Cooling water will be evaporated or stored in ponds, which will be used to recycle the water as much as possible. Neither of these processes contributes to groundwater contamination.

Drawdown of the regional water table to the degree that it will influence local water wells is not expected. There are abundant supplies of groundwater in the Ingham and Jackson areas. Groundwater withdrawals will not be centered in one area. The pumping distribution around the SSC ring will be distributed in such a manner that no impact on local water supplies should be felt. In the event that there is a problem at a select area, the State of Michigan's commitment to replace water supplies affected by the SSC is still valid. [Heinzman 1988c]

12.1.7 What happens if wells go dry? Will new ones be drilled? Who will do it? How many years will it take? [Harshbarger 1987]

If a water well is temporarily affected by dewatering, an interim water supply will be made available immediately. When conditions are such that a new well is necessary it will be replaced. The State of Michigan has taken the responsibility to supply water during construction or other activities that may temporarily disrupt water use in an area. The time scale is immediately with respect to the replacement of wells, not years. [Heinzman 1988c]

12.1.8 How far away will our wells have to be from the tunnel? Approximately 100 feet from the tunnel. [Q Sheet 2/16/88]

The DOE has requested that no wells be placed any closer than 150 feet from the tunnel. This means that a zone 300 feet wide along the tunnel will not have water wells located in it. The replacement of these water supplies are covered by the State of Michigan's commitment to replace water supplies that are affected by the SSC construction. [Heinzman 1988c]

12.1.9 If city water has to be brought to rural people, will it be metered? [Harshbarger 1987]

The answer to this question is not known at this time. There is a possibility that all individuals hooked up to a city water system would be required to have meters installed. [Heinzman 1988c]

12.1.10 Who will maintain the two county drains going through the campus area? Will the county drain commissioner be involved to assure proper water discharge? [Hot Line 1988a]

The maintenance of the drains should continue to be the responsibility of the County Drain Commission. This is consistent with what the DOE has said about honoring state and local codes with respect to operation and maintenance of the facility. [Heinzman 1988c]

BIBLIOGRAPHY

Ann Arbor News
1988 "The nuts and bolts of the Super Collider," by Max Gates, March 15, p. B2.

Birch, Alan, Gilbert/Commonwealth Inc.
1988 Personal Communication, March.

Burdock, Jack, Michigan SSC Commission
1988a Personal Communication, Feb. 15.
1988b Written response to questions, April 18.
1988c Personal Communication, April 26.

Caulder, B.J.
1977 "Focus Groups and the Nature of Qualitative Marketing Research," *Journal of Marketing Research*, Vol.14, pp. 353-364.

European Center for Nuclear Research (CERN)
1987 "Allgemeine Angaben," p. 2.

Fermi National Accelerator Laboratory
1988 Personal Communication.

Gilbert/Commonwealth Inc.
1987 *SSC Information Sheet, Answers to Frequently Asked Questions about the SSC.*

1986 *Southeast Michigan Superconducting Super Collider Feasibility Study and Preliminary Site Evaluation*, Jackson Michigan.

Grobe, Edward, director, Ingham County Department of Development
1988 Questions mentioned at SSC Study Team meeting.

Grover, Michael, SSC Community Representative
1988 Personal Communication, April 18.

Harshbarger, Camilla L., Richard W. Stoffle, Florence V. Jensen, and Michael W. Traugott
1987 *Social Assessment of the Superconducting Super Collider in Michigan: Public Meeting and Focus Group Questions*, Ann Arbor, Michigan: Institute for Social Research, University of Michigan.

Heinzman, James, Michigan SSC Commission, geologist
1988a Personal Communication, Feb. 16.
1988b Update at SSC Study Team meeting, April 18.
1988c Written response to questions, April 26.

Hohl, Susan, Michigan Department of Transportation, Real Estate Division
1988 Written response to questions, March 25.

Hot Line, for the SSC project in Michigan
1988a Feb. 9.
1988b March 3.

Jackson, J. David, Maury Tigner and Stanley Wojcicki
1988 "The Superconducting Supercollider," *Scientific American* Vol. 254, Number 3, pp. 66-77.

Jones, Lawrence W., SSC Central Design Group; The University of Michigan, Professor of Physics
1988 Written response to questions, March 3.

1986 "Radiation Safety of the Superconducting Super Collider," Univeristy of Michigan.

Kane, Gordon, The University of Michigan, Professor of Physics
1988a Personal Communication, Feb. 11.
1988b Personal Communication, April 25.

Limon, Peter, SSC Central Design Group,
1988 University of Michigan Physics Colloquium, Feb. 10.

Lowi, Theodore J., and Benjamin Ginsberg
1976 *Policide*, New York: MacMillan Publishers, Inc.

Michigan Department of Transportation
nd *Your Rights and Benefits as a Person Displaced From a Transpo Project*, Lansing, Michigan.

Michigan SSC Study Team
1988a "Real Estate Acquisition Schedule," Lansing, Michigan.

1988b "Making People Whole," Lansing, Michigan.

Morgan, David L. and Margaret T. Spanish
1984 "Focus Groups: A New Tool for Qualitative Research," *Qualitative Sociology*, Vol 7, pp. 253-269.

National Academy of Sciences and National Academy of Engineering
1988 "Siting the Superconducting Super Collider," National Academy Press: Washington D.C.

Ott, Steve, Gilbert/Commonwealth Inc.
1988 Personal Communication, Feb. 16.

Question sheets recorded at Department of Energy meeting on Feb. 16, 1988.

Radioactivation Task Force
1987 "Report of the Task Force on Radioactivation."

Response Cards filled out by public at Department of Energy meeting on Feb. 16, 1988.

Rubenstein, James M. and Robert S. Bacon
1983 *The Cultural Landscape: An Introduction to Human Geography*, West Publishing Company: St. Paul.

Salgado, United States Department of Energy
1988 Memorandum, Attachment Two, p. 3.

SSC Central Design Group
1987 "An Introduction to Radiation Protection for the Superconducting Super Collider," Katharine Metropolis, editor.

1986 "Conceptual Design of the Superconducting Super Collider."

SSC Tape Manuscripts, The University of Michigan, Institute for Social Research:
1988 White Oak Public Meeting, Jan. 25
1988 Stockbridge Public Meeting, Jan. 26.

State of Michigan
1987a *Siting the Superconducting Super Collider in Michigan: Stockbridge Site*. Lansing, Michigan: Office of the Governor.

1987b *Siting the Superconducting Super Collider in Michigan: Dundee Site*. Lansing, Michigan: Office of the Governor.

Stoffle, Richard W., The University of Michigan, Institute for Social Research
1988 Personal Communication, Feb. 16.

Stoffle, Richard W. & Michael W. Traugott.
1986 "Social Impact Assessment of the Michigan Superconducting Super Collider: A Scoping Study Proposal."

Stoffle, Richard W., Michael W. Traugott, Camilla L. Harshbarger, Florence V. Jensen and Michael J. Evans
1988 *Assessing Community Support for the Superconducting Super Collider at the Stockbridge, Michigan Site: A Report to the United States Department of Energy*, Ann Arbor, Michigan: Institute for Social Research, University of Michigan, Feb. 16.

Stoffle, Richard W., Michael Traugott, Florence V. Jensen, and Robert Copeland

1987 *Social Assessment of High Technology: The Superconducting Super Collider in Southeast Michigan*. Ann Arbor, Michigan: Institute for Social Research, University of Michigan.

Super Collider Site Evaluation Committee, National Academy of Sciences and National Academy of Engineering

1988 *Siting the Superconducting Super Collider*, Washington D.C.: National Academy Press.

Town Crier

1988 "'Losses few; benefits enormous' says Warrenville mayor on SSC," Stockbridge, Michigan: The Town Crier, March 1.

Traugott, Michael W., The University of Michigan, Institute for Social Research

1988 Personal Communication, Feb. 15.

United States Department of Energy

1987a *Invitation for Site Proposals - Superconducting Super Collider*, Washington D.C.: U.S. Department of Energy Research, SSC Site Task Force.

1987b *Superconducting Super Collider*, Washington D.C.: U.S. Department of Energy.

1987c *Waste Generation at the Superconducting Super Collider*, Washington D.C.: U.S. Department of Energy.

nd *The SSC, The Project and the Environmental Process*, Washington D.C.: U.S. Department of Energy.

Universities Research Association

nd *The Superconducting Super Collider*, Washington, D.C.

1987 *To the Heart of Matter - The Superconducting Super Collider*, Washington D.C.: Universities Research Association.

1985 "Super Collider Research & Development, The First Two Years." Washington, D.C.

Wall Street Journal
1988 "Two new superconductors provide clue to making such materials commercial," by David Stripp, March 23.

Wagner, P.L.
1962 *Readings in Cultural Geography*, The University of Chicago Press: Chicago and London.